Valerio Voliani

Gold Nanoparticles

Also of Interest

Chemistry of Nanomaterials.
Volume 1 A: Metallic Nanomaterials
Kumar (Ed.), 2018
ISBN 978-3-11-034003-7, e-ISBN 978-3-11-034510-0,
e-ISBN (EPUB) 978-3-11-038382-9

Chemistry of Nanomaterials.
Volume 1 B: Metallic Nanomaterials
Kumar (Ed.), 2018
ISBN 978-3-11-063660-4, e-ISBN 978-3-11-063666-6,
e-ISBN (EPUB) 978-3-11-063698-7

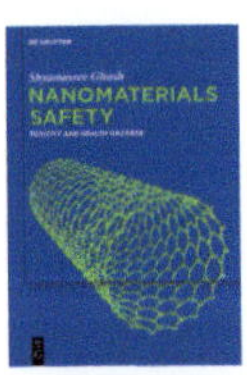

Nanomaterials Safety.
Toxicity And Health Hazards
Ghosh, 2018
ISBN 978-3-11-057808-9, e-ISBN 978-3-11-057909-3,
e-ISBN (EPUB) 978-3-11-057836-2

Nanomaterials in Joining.
Charitidis (Ed.), 2015
ISBN 978-3-11-033960-4, e-ISBN 978-3-11-033972-7,
e-ISBN (EPUB) 978-3-11-038950-0

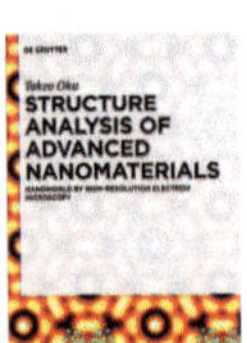

Structure Analysis of Advanced Nanomaterials.
Nanoworld by High-Resolution Electron Microscopy
Oku, 2014
ISBN 978-3-11-030472-5, e-ISBN 978-3-11-030501-2,
e-ISBN (EPUB) 978-3-11-038804-6

Valerio Voliani

Gold Nanoparticles

An Introduction to Synthesis, Properties
and Applications

DE GRUYTER

Author
Dr. Valerio Voliani
Istituto Italiano di Tecnologia
Cente for Nanotech. Innovation @NEST
Piazza San Silvestro 12
56127 Pisa
Italy

ISBN 978-1-5015-1901-7
e-ISBN (PDF) 978-1-5015-1145-5
e-ISBN (EPUB) 978-1-5015-1157-8

Library of Congress Control Number: 2020930376

Bibliographic information published by the Deutsche Nationalbibliothek
The Deutsche Nationalbibliothek lists this publication in the Deutsche Nationalbibliografie; detailed bibliographic data are available on the Internet at http://dnb.dnb.de.

Cover image: Manudri / iStock / Getty Images Plus
Typesetting: Integra Software Services Pvt. Ltd.
Printing and binding: CPI books GmbH, Leck

www.degruyter.com

Preface

Gold.
Gold is the most ancient known element.
Gold is a bright, yellow, dense, soft, malleable, and ductile metal.
Gold is an element that has fascinated humans since its advent.
Gold origins are celestial.
Gold is simply, gold.

This book has just one protagonist, gold. Back in time, humanity has always woven its history with gold. Never before, however, it has been such precious for humans. Indeed, several next treatments for various diseases will be based on nanostructured gold.

In Chapter 1, the mostly used methods for the *wet chemistry* synthesis of gold nanostructures, among which spheres, cages, cubes, and rods are described, and an exhaustive "guide to the synthesis" has been reported. In Chapter 2, the optical and physiological behaviours of gold nanoparticles have been comprehensively discussed and the most common encapsulation processes explained. Finally, in Chapter 3 the most recent advances in the field of nanomedicine are reported. All chapters comprise a complete and recent bibliography, in order to give to the readers the opportunity to further extend the addressed topics.

This book is a unique guide on gold nanoparticles from the synthesis to the currently most promising applications in clinical setting. Overall, thanks to this book, the general reader can achieve a broad vision on nanostructured gold, while physicians and biologists can have a look to the new breakthroughs in medical field, and chemists have a general guide to the synthesis of gold nanoparticles.

Valerio Voliani
Pisa, Italy
November, 2019

https://doi.org/10.1515/9781501511455-202

Introduction

Gold is a relatively rare element and its name may be derived from Proto-Germanic **gulþą* from Proto-Indo-European **ǵʰelh₃-* ("to shine, to gleam; to be yellow or green"). Its symbol in the periodic table of the element is Au (from Latin: aurum) and has one of the higher atomic numbers of the transition metals: 79. It seems eternal as it is resistant to oxygen or acids due to its impressive stability, and this property has determined the huge appeal it has always had on humanity. Indeed, it has been widely employed from ancient times in religious practices as well as currency throughout the world in efficient indirect exchange with respect to barter, and to store wealth in hoards. For exchange purposes, mints produce standardized gold bullion coins, bars, and other units of fixed weight and purity. In some Semitic religions, gold has been associated to both holiness and evil. In the Book of Exodus, the Golden Calf is a symbol of idolatry, while in the Book of Genesis, Abraham was said to be rich in gold and silver, and Moses was instructed to cover the Mercy Seat of the Ark of the Covenant with pure gold. In Byzantine iconography, the halos of Christ, Mary, and the Christian saints are often golden. Gold resulted so important for humanity that its production by transmutation has long been a subject of inquiry, resulting in the foundation of the discipline of alchemy. Curiously, it was really produced from other metals in 1924 by the Japanese physicist Hantaro Nagaoka, who synthesized gold from mercury by neutron bombardment. In 2017, China was the world's largest gold producer by extraction (440 tonnes), followed by Australia, (300 tonnes) and Russia (255 tonnes). Interestingly: (i) 75% of the presently accounted for gold has been extracted in the last century, (ii) the currently known amount of gold would form a single cube 20 m on a side (8,000 m^3), and (iii) in Japan the 16% of gold is used in electronic technology.

Gold (intended as metal and not ion) is non-toxic and non-irritating when ingested. On this hand, gold is approved as a food additive in the EU (additive E175) and it is employed in some food preparation (as Varak) and as a component of alcoholic drinks (Goldschläger, Gold Strike, and Goldwasser). Furthermore, due to its employment by shamanic practitioners, it seems the most anciently administered medicine. In medieval period, gold was often perceived as beneficial for the health, while in the 19th century gold has been employed for the treatment of amenorrhea, impotence, alcoholism, and nervous disorders, among which depression, epilepsy, and migraine. Nowadays, some gold salts (sodium aurothiomalate and auranofin) are employed in the treatment of arthritis due to their anti-inflammatory properties that reduce the pain and swelling of rheumatoid arthritis.

In the last two decades, gold in medicine has become more and more precious in its metal form while reducing its size. The first contacts between humans and nanostructured gold happened during the manufacture of stained glass in ancient times, when ruby red glasses were obtained by adding gold salts during the production process. More recently, in 1857, Michael Faraday intentionally produced colloidal

https://doi.org/10.1515/9781501511455-203

gold for his electrochemical experiments and discovered that the optical properties differing from those of the corresponding bulk metal. Nowadays, colloidal gold is essential for the development of innovative or combined/enhanced treatments for several diseases. Indeed, nanostructured gold can be exploited for its peculiar behaviors, among which localized surface plasmon resonance(s) (LSPRs), X-rays attenuation, and extreme stability/biocompatibility. Remarkably, these features add up to the other breakthroughs carried by organic nanomaterials, such as the high drug- loading efficiency and the localized action, that have already resulted in a revolution during the 1990'ies for the clinical treatment of several neoplasms.

Noticeably, despite the massive efforts during the last years, treatments based on colloidal gold are still at the preclinical stage, due to the body persistence issue. Indeed, the non-biodegradable nature of gold results in longstanding persistence in some organs, among which the liver. On the other hand, therapeutics have not to leave residues in patients after the action, as suggested by all the approbation agencies. Very recently, new breakthroughs have emerged to overcome this drawback, among which the ultrasmall-in-nano approach, bringing again gold colloid to the forefront of clinical applications.

It is worth to remember several other critical challenges that have to be dealt with to promote the translation of colloidal gold in clinical settings, such as the standardization of the materials/protocols and the scaling-up/automatization of the productions.

In summary, gold is more than the metal that humans known from ancient times, and today it can be useful as never before.

Valerio Voliani
Pisa, Italy
November, 2019

Contents

1 Synthesis of gold nanostructures

Optical and biological properties of noble metal colloids result from a delicate balance between material, size, shape, and dispersion of the nanostructures (see Chapter 2) [1]. The interest in tuning the behaviors of nanomaterials has encouraged the exploration of many synthetic processes in the last two decades, in order to produce nanoparticles with peculiar features set for specific applications [2]. In this chapter, the most common protocols to produce gold nanoparticles (AuNPs) are discussed and a practical guide to the synthetic processes is reported at the end of the chapter. In this new edition, a section on gold–silica hybrid nanomaterials has been added due to their potentiality for the clinical translation of noble metals to the clinical practice [3].

In general, nanostructures are produced by two approaches: bottom-up and top-down [4]. In the first approach, the synthesis begins from the interaction of metal ions, in order to build a more complex structure, while in the latter the bulk material is gradually eroded by physic-chemical mechanisms until the desired size and shape is achieved. Nanostructures for biological applications are usually produced by employing the bottom-up approach, in which the lower control over the final product is balanced by the amount/batch [5]. In the bottom-up approach, synthesis are based on the reduction of salts of the metal of interest (the precursor) in the presence of reducing and surfactant agents in aqueous or organic media. By changing some key variables, among which the reactants, their relative molar concentrations, the reaction temperature, or the stirring velocity, the nucleation and growth processes can be controlled, achieving colloids with the desired properties (Figure 1.1).

The control on the temporal separation of these two effects is critical to obtain AuNPs with a narrow size-dispersion (Figure 1.1). In the following, the synthetic strategies are named as usually accepted.

1.1 Gold nanospheres (AuNSs)

1.1.1 Turkevich method

The method proposed by Turkevich is based on the reduction of tetrachloroauric acid ($HAuCl_4$) with sodium citrate in water at 90–100 °C [7]. This is the most common process to synthesize AuNSs due to its simple and environmental benign procedure. Furthermore, the size of the nanospheres can be readily tuned from 10 to 150 nm by varying the molar ratio of citrate to $HAuCl_4$. The experimental protocols are based on a rapid addition of sodium citrate solution to a hot (90–100 °C), aqueous solution of $HAuCl_4$. In the redox reaction, sodium citrate acts both as a reducing/capping and buffering agent. In particular, a study by Ji and coworkers

https://doi.org/10.1515/9781501511455-001

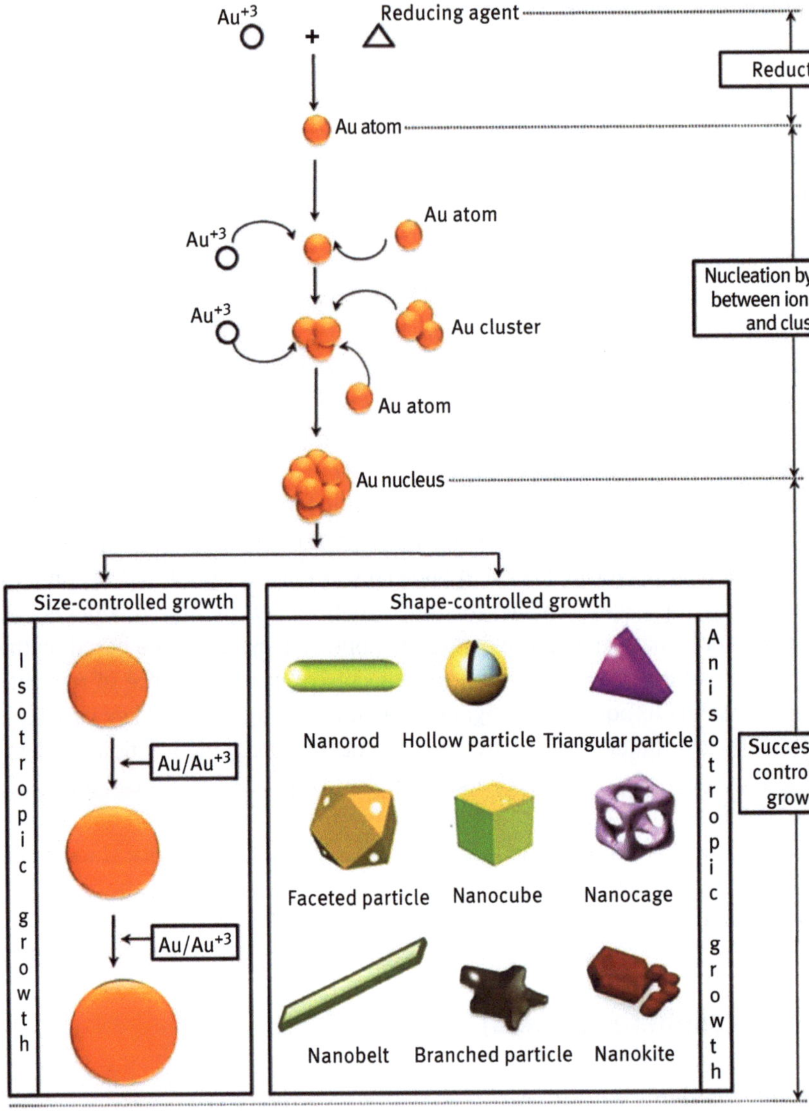

Figure 1.1: Formation mechanism of gold nanoparticles (AuNPs) with various particle sizes and shapes by chemical reduction method. Reproduced with permission from D.T. Nguyen, D.J. Kim and K.S. Kim, Micron, 2011, 42, 207. ©2011, Elsevier [6].

highlighted the fact that citrate species buffer the pH of the $HAuCl_4$ solution from pH ≈ 2 to higher values (even neutral), depending on the amount of citrate added [8]. Accordingly to the pH of the reaction solution, $AuCl_4$- ions [acid dissociation constant (pK_a) 3.3] are hydrolyzed to several kind of auric precursor ions, such as:

$AuCl_3(OH)^-$ (pK_a 6.2), $AuCl_2(OH)_2$- (pK_a 7.1), $AuCl(OH)_3$- (pK_a 8.1), and $Au(OH)_4$- (pK_a 12.9). Their reactivity decreases as in the following sequence: $AuCl_4$- > $AuCl_3(OH)^-$ > $AuCl_2(OH)_2$- > $AuCl(OH)_3$- > $Au(OH)_4$- [7, 9]. In the first step of the reaction (Figure 1.2), sodium citrate is oxidized to sodium acetone dicarboxylate (SADC) while any precursor of $HAuCl_4$ is reduced to AuCl by pH-dependent kinetics.

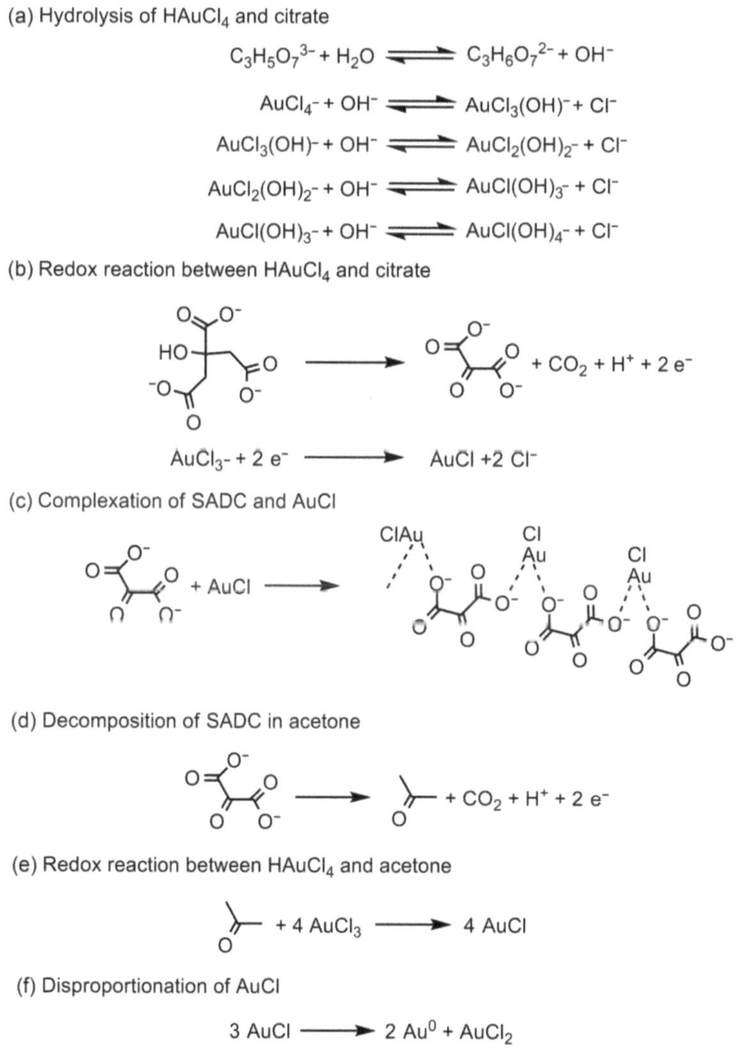

Figure 1.2: Hypothesis of the reactions involved in the formation of colloidal gold in the citrate methods. Reproduced with permission from H.B. Xia, S.O. Bai, J. Hartmann and D.Y. Wang, Langmuir, 2010, 26, 3585. ©2010, American Chemical Society [10].

The nucleation process (formation of gold clusters) is caused by the formation of AuCl/SADC macromolecular complexes and, subsequently, the growing steps to AuNSs are catalyzed by gold clusters themselves, which support the dismutation of AuCl on their surfaces (Figure 1.2) [11]. It is important to remember here that a dismutation reaction occurs when the same ion acts as oxidant and reducing agent. The coagulation of macromolecular SADC/AuCl complexes is induced by concentration fluctuation in agreement with LaMer theory [12]. Matijevic and coworkers have demonstrated that rapid coagulation favors the formation of monodisperse spherical particles [13]. Thus, a rapid formation of a large amount of SADS support the formation of nanoparticles with a narrower size distribution. On the other hand, SADC can also readily decompose to acetone at high temperatures (>90 °C), especially at neutral to basic pH [14]. Acetone can reduce the $HAuCl_4$ precursor ions to AuCl, leading to a secondary nucleation that results in a broadening of the size distribution of the colloid. The secondary nucleation can be decreased by fast oxidation of citrate to SADC at a lower pH using silver ions under ultraviolet (UV)-light irradiation [15]. Interestingly, silver ions also foster the formation of very spherical nanostructures (Figure 1.3(c)). For example, Xia and colleagues [9] improved the *Turkevich method* by adding a catalytic quantity of Ag^+ ions during the reaction, obtaining quasispherical nanoparticles [10]. Theoretical calculations have demonstrated that different gold facets caused peculiar deposition potential for Ag: 0.12, 0.17, and 0.28 eV for (111), (100), and (110) facets, respectively [16]. This suggests that Ag atoms produced by citrate reduction of Ag^+ ions deposit preferentially on the (110) and (100) facets of AuNSs. Then, the Ag layer is oxidized and replaced by gold ions. Overall, the deposition of silver significantly decreases the growth rate of AuNSs on the (110) and (100) facets, promoting the formation of spheres. The presence of weak ligands such as citrate does not interfere with the deposition and decomposition of Ag^+ on AuNSs, enabling the reshaping of the polycrystalline nanoparticles to quasispherical geometries [11].

In summary, because of the many interdependent variables involved in the reaction process, AuNSs obtained by the *Turkevich method* (Figure 1.3(a)) usually have a broad distribution of size and shape. Indeed, considering that both the nucleation and crystal growth of AuNSs are very fast at high temperatures (less than 10 min at 100 °C), the buffering effect of citrate, and the inhomogeneous nucleation caused by potential nonuniform mixing, a temporal overlap between nucleation and crystal growth may happen, broadening the final size distribution of the colloid. For example, despite the several improvements to the *Turkevich method*, AuNSs are generally produced with a diameter in the range of 12–60 nm with a relative size distribution of 13%–16% and usually with a nonuniform and irregular shape (such as quasispheres, ellipsoids, and triangles) [10, 17–19]. On the other hand, it is useful to remember that if geometrical uniformity is not a key-point for the investigation or the final application, the *Turkevich method* is the most easy and fast process to produce high throughput water-soluble AuNSs.

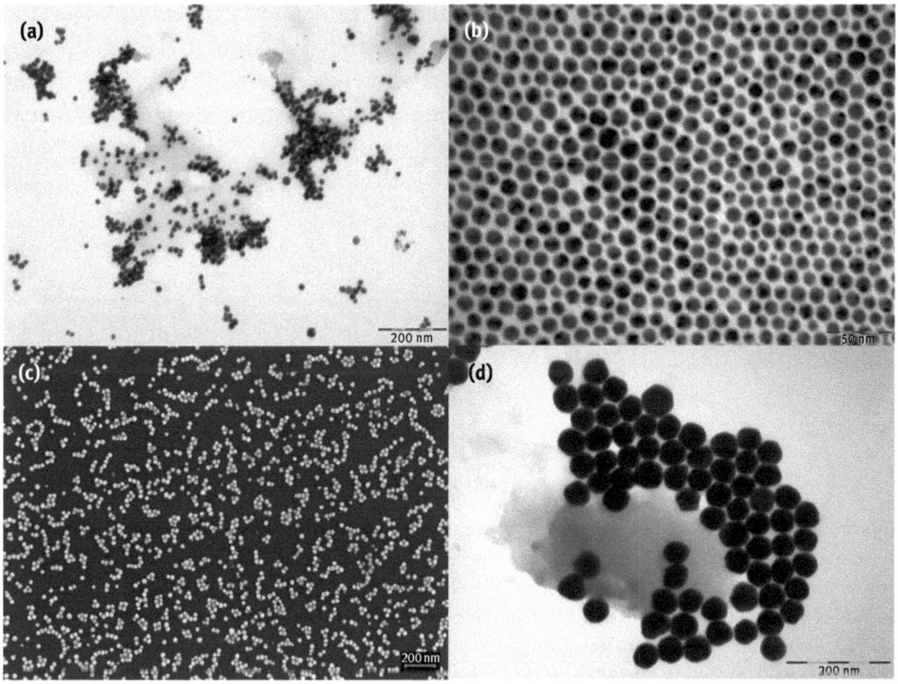

Figure 1.3: (a) Transmission electron microscopy (TEM) image of gold nanospheres synthesized by the *Turkevich method*. Average sphere diameter of 20 nm, scale bar 200 nm. (b) TEM image of gold nanospheres synthesized by the *Brust method*. Average sphere diameter of 10 nm, scale bar 50 nm. (c) Scanning electron microscopy (SEM) image of gold nanospheres synthesized by the *Xia method*. Average sphere diameter of 15 nm, scale bar 200 nm. (d) TEM images of gold nanospheres synthesized by the *Zhong method*. Average sphere diameter of 30 nm, scale bar 200 nm.

1.1.2 Zhong method

The *Zhong method* yields water-soluble AuNPs with size in the range of 10–100 nm diameter and dispersion down to 2–5% (Figure 1.3(d)) [20]. It is a two-step reaction that involves the synthesis of gold seeds followed by a seeded growth process in aqueous solutions using sodium acrylate as a reducing and capping agent. In the first step, like the *Turkevich method*, $HAuCl_4$ gold precursors are reduced by sodium acrylate at 100 °C, yielding 15 nm diameter gold seeds. The major by-products of the oxidation of sodium acrylate (Wacker reaction) are CH_3COCO_2H (pyruvic acid or 2-oxopropanoic acid) and $CHOCH_2CO_2H$ (3-oxopropanoic acid) [21]. The second step involves a seeded and aggregative growth mechanism, as demonstrated by Njoki and coworkers [22]. The growth of larger and monodisperse particles arise from the reduction of Au(III) on the surface of Au seeds by a mix of acrylic acid and sodium acrylate agents. The reaction occurs at room temperature (20 ± 0.5 °C) in about three days under control of the pH and the reaction temperature. The size-dispersion and the shape-uniformity are controlled by the Ostwald ripening (Figure 1.4), a mechanism often considered to be the general driving force for the growth of particles. In the Ostwald ripening, smaller particles dissolve and recrystallize onto larger particles, triggering the growth of larger particles from smaller ones [23].

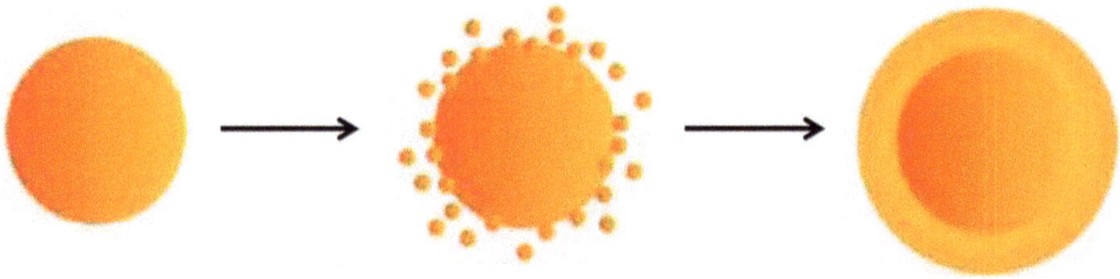

Figure 1.4: Illustration of the seeded and aggregative growth mechanism for the controlled growth of gold nanoparticles. In the central picture the coalescence of the smaller particles on the larger one is shown. Reproduced with permission of P.N. Njoki, I-I.S. Lim, D. Mott, H-Y. Park, B. Khan, S. Mishra, R. Sujakumar, J. Luo and C-J. Zhong, Journal of Physical Chemistry C, 2007, 111, 14664. ©2007, American Chemical Society [20].

Using a spherical model for AuNSs with an initial seed radius r (cm) and a growth thickness d (cm), in order to produce particles of total radius $r + d$ (cm), the mass balance between the grown seeds and the concentration C of $AuCl_4$- (mM) is shown in eq. (1.1):

$$N \times \left[\frac{4\pi}{3}(r+d)^3 - \frac{4\pi}{3}r^3\right] \times \rho = C \times V \times M \quad (1.1)$$

where N is the number of AuNPs in the total volume, ρ is the density of bulk gold (18.9 g/cm^3), M is the molecular weight of Au (197 g/mol), C is the concentration of Au precursor (mol/l), and V is the total reaction volume (L) [24]. From eq. (1.1) is derived the d/r expression of eq. (1.2):

$$\frac{d}{r} = \sqrt[3]{1 + 2.49 \times \frac{C \times V}{N \times r^3}} - 1 \quad (1.2)$$

The desired monodispersed AuNSs can be achieved by using the eq. (1.2) to finely control the molar ratio of the reactants. In summary, the *Zhong method* shows some peculiar interesting features: (i) very narrow size-dispersion, (ii) high throughput production, (iii) reproducibility, and (iv) good control over the geometry of AuNSs. On the other hand, the reaction is time-consuming and the success of the process is strictly related to the control of the reaction temperature.

1.1.3 Brust method

The strategy proposed by Brust (*Brust method*) consists of growing the metallic nanospheres with the simultaneous attachment of self-assembled thiol monolayers on the growing nuclei [25]. In order to allow the surface reaction during metal nucleation and growth, the particles are produced in a two-phase system. In this method, $AuCl_4$- is transferred from an aqueous solution to toluene using tetraoctylammonium bromide as the phase-transfer reagent and reduced with aqueous sodium borohydride in the presence of dodecanethiol ($C_{12}H_{25}SH$). Upon addition of the reducing agent, the organic phase changes color from orange to deep brown within a few seconds. The overall reaction is summarized in eq. (1.3):

$$\begin{aligned} &AuCl_4\text{-}_{(aq)} + N(C_8H_{17})_4 + (C_6H_5Me) \rightarrow N(C_8H_{17})_4 + AuCl_4\text{-}(C_6H_5Me) \\ &mAuCl_4\text{-}(C_6H_5Me) + nC_{12}H_{25}SH(C_6H_5Me) + 3\,me^- \rightarrow 4\,mCl^-_{(aq)} + \\ &(Au_m)(C_{12}H_{25}SH)_n(C_6H_5Me) \end{aligned} \tag{1.3}$$

where the source of electrons is BH_4-. The result of the reaction is determined by the ratio of thiol to gold, that is, the ratio *n/m*. This single reaction yields a surface-functionalized gold colloid in the range of 2–8 nm diameter with a dispersion of about 4–6% (Figure 1.3(b)). The kinetics of AuNSs growth is determined by the thiol surface coverage, and, as usual in two-phase reaction, AuNSs size can be mainly controlled by the reaction conditions. The major drawback of this reaction relies on the difficulty of growing uniform and regular-shape nanostructures with diameter >25 nm. Also, this reaction is performed in organic media, resulting in AuNSs that requires further steps to obtain water soluble nanoparticles useful for bioapplications [26].

1.2 Gold nanoparticles

1.2.1 Gold Nanorods

The first approach to the synthesis of gold nanorods, based on an electrochemical route within reverse micelles, was introduced by Wang and coworkers (Figure 1.5) [27].

The synthesis is carried out in an electrochemical cell with two 3.0 cm × 1.0 cm × 0.05 cm electrodes: the sacrificial anode is formed by a plate of gold, and the cathode is formed by a plate of platinum. The electrodes are immersed in an electrolytic solution containing a mix of two cationic surfactants, hexadecyltrimethylammonium bromide (CTAB), and a small amount of a more hydrophobic cationic surfactant, tetradodecylammonium bromide, which acts as a rod-inducing cosurfactant. CTAB is used to both support the electrolyte solution and as a stabilizer for the final colloid, in order to prevent aggregation phenomena. An appropriate amount of

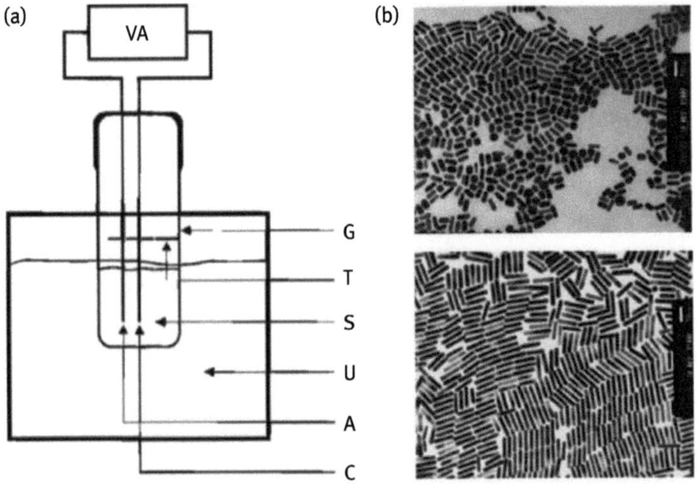

Figure 1.5: (a) Schematic diagram of the set-up for preparation of gold nanorods via the electrochemical method containing: (i) VA (volt-ampere), power supply, (ii) G, glass electrochemical cell, (iii) T, Teflon spacer, (iv) S, electrode holder, (v) U, ultrasonic cleaner, (vi) A, anode, (vii) C, cathode. (b) TEM micrographs of Au nanorods with different aspect ratios: 2.7 (top) and 6.1 (bottom). Scale bars represent 50 nm. Reproduced with permission of J. Pérez-Juste, I. Pastoriza-Santos, L.M. Liz-Marzán and P. Mulvaney, Coordination Chemistry Reviews, 2005, 249, 1870. ©2005, Elsevier [28].

acetone and cyclohexane are usually added to the electrolytic solution in order to promote the inclusion of the cosurfactant into the CTAB micelles, which support their elongation [29]. A typical electrolysis process (Figure 1.5) is conducted at a current intensity of 3 mA for 30′. During the synthesis, the anode in gold is consumed, to form $AuBr_4$-. Anions are complexed by the cationic surfactants and migrate to the cathode where the reduction to metal gold nanorods occurs. At present, it is unclear whether nucleation occurs on the cathode surface or within the micelles. A key factor to control the aspect ratio of the Au nanorods is the presence of a silver plate inside the electrolytic solution, which is gradually immersed behind the cathode. The redox reaction between gold ions generated from the anode and silver metal leads to the formation of silver ions. Wang and coworkers have found that the concentration of silver ions and their release rate control the length of the nanorods [27]. The complete mechanism, as well as the role of the silver ions, is still unknown.

Nowadays, the *seeded growth method* is the most used process to synthesize gold nanorods in water. In this two-step method, the primary nuclei are prepared by borohydride reduction of $HAuCl_4$ in the presence of CTAB as capping agent, in order to form 3.5 nm diameter gold seeds [30]. In addition, the growing step is carried out in aqueous media and in presence of CTAB. It is worth to notice that potential secondary nucleation during the growing stage is inhibited by carefully controlling the growth conditions, and in particular by using ascorbic acid as a weak reducing agent. Indeed, ascorbic acid cannot reduce gold salts in the presence of micelles if gold seeds are not present. Interestingly, the aspect ratio of gold nanorods is easily tunable by controlling the growth conditions (Figure 1.6).

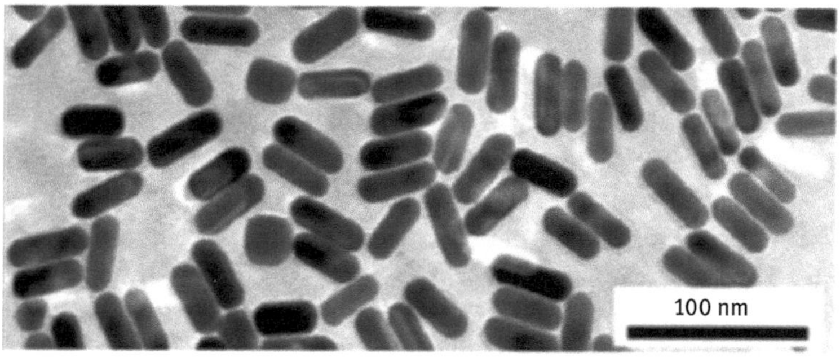

Figure 1.6: TEM images of gold nanorods indicate an average length and width of 48.1 ± 5.5 nm and 14.3 ± 2.2 nm, respectively. Reproduced with permission of N.J. Durr, T. Larson, D.K. Smith, B.A. Korgel, K. Sokolov and A. Ben-Yakar, Nano Letters, 2007, 7, 941. ©2007, American Chemical Society [31].

Aspect ratio as well as the monodispersity and reaction yield are dependent on several factors, among which the stability of the seed, the reaction temperature, the conductance of the water, and the kind/concentration of the surfactant. Furthermore, the addition of $AgNO_3$ to the growing solution influences the mechanism of nanostructure formation, and thus the yield, the aspect ratio, and also the crystal structure of the resulting gold nanorods [28]. In this regard, a very interesting guide on “how to produce the desired nanorods” comprehensive of many tips and tricks was recently reported by Scarabelli et al. [32].

The mechanism of formation of rod-shaped nanoparticles in aqueous surfactant media still remains mostly unclear [28]. Based on the idea that CTAB absorbs onto gold nanorods in a bilayer moiety, with the trimethylammonium head-groups of the first monolayer on the surface of the nanoparticle, it was proposed that the CTAB head-group preferentially binds to the crystallographic faces of the sides of the rods rather than the faces at the tips [33, 34]. Thus, the syntheses of gold nanorods are governed by preferential adsorption of CTAB to different crystal faces during the growth, rather than acting as a soft micellar template [34].

Gao and coworkers reported on the geometrical dependence of the rods from CTAB analogues, in which the length of the hydrocarbon tails was varied keeping the head-group and the counterion constant [35]. The length of the surfactant tail resulted critical to control the length of the nanorods and the reaction yield, that is, the shorter chain lengths produce shorter nanorods in low yields and vice versa. The seeds are another fundamental parameter to take into account. For example, the yield of nanorods prepared by CTAB capped seeds is higher than from citrate-stabilized seeds. Moreover, the lower the amount of seed added, the higher the aspect ratio of the nanorods formed. It is important to remember here that an increase in the aspect ratio of nanorods implies a red-shift of the localized surface plasmon resonances (LSPRs). In summary, the general observations over gold nanorods production are as follows [28]: (i) the yield of rods improves with increasing colloidal stability of the seeds, thus dimers or aggregates are not precursors of rods, (ii) the length of the surfactant tail increases the yield and the aspect ratio of the nanorods, (iii) an increase in the ionic strength produces a decrease in the yield of rods, (iv) the aspect ratio can be controlled through seeds to $HAuCl_4$ ratio in the growing solution, that is, an increase in the amount of seeds solution means a decrease in the aspect ratio of the rods, (v) both $AuCl_4$- and $AuCl_2$- are quantitatively adsorbed to CTAB [36], and (vi) the higher the curvature of the gold surface is, the faster is the rate of growth. The redox reactions between ascorbic acid and $HAuCl_4$ in the presence of CTAB (and the subsequent dismutation of Au^+) can be described by eqs. (1.4)–(1.7):

$$AuCl_4{-} + 2\,e^- \leftrightarrow AuCl_2{-} + 2\,Cl^- \tag{1.4}$$

$$3\,AuCl_2{-} \leftrightarrow AuCl_4{-} + 2\,Au^0 + 2\,Cl^- \tag{1.5}$$

$$AuCl_2{-} + e^- \leftrightarrow Au^0 + 2\,Cl^- \tag{1.6}$$

$$AuCl_2{-}(CTAB) + Au_m \leftrightarrow Au_{m+1}(CTAB) + CTAB + 2\,Cl^- \tag{1.7}$$

Ascorbic acid reduces Au(III) to Au(I) in the presence of CTAB (eq. (1.4)). However, the dismutation of $AuCl_2$- does not occur (eq. (1.5)) in the absence of gold seeds. Consequently, the reduction of Au(I) should proceed through electron transfer at the surface of the electron-rich gold seeds (eq. (1.6)). Then the total redox process can be described by eq. (1.7).

The presence of silver nitrate allows better control of the shape of gold nanorods, even if the mechanism by which Ag^+ ions modify the metal nanoparticle shape is not fully understood. It has been hypothesized that Ag^+ adsorbs on the particle surface in the form of AgBr (thanks to the bromide species related to CTAB) and hinders the growth of the AgBr-passivated crystal facets [37]. Moreover, the reduction process of silver ions due to ascorbic acid is prevented due to the experimental conditions (pH~2.8 and room temperature) [38]. However, there is a critical silver ion concentration, above which the aspect ratio of the nanorods decreases again [33].

1.2.2 Gold nanocubes and polyhedral nanocrystals

Despite their very interesting behavior, only few synthetic processes were explored for the synthesis of gold nanocrystals with many edges in the size range of 20–80 nm [39]. The two main reactions are proposed by Wu and coworkers and by Seo and coworkers [40, 41].

The method proposed by Wu and coworkers is a versatile process similar to the synthesis of gold nanorods. The process is, such as in the case of rods, a two-step seeding growth method, and the main reactants are a gold precursor ($HAuCl_4$), ascorbic acid as reducing agent, and, as usual, hexadecyltrimethylammonium chloride (CTAC) as a capping agent. By changing the reactants' concentrations (Figure 1.7), they were able to produce nanocrystals with a well-controlled size and morphology: cubes, bipyramids, truncated cubes, stars, and rhombic dodecahedrons.

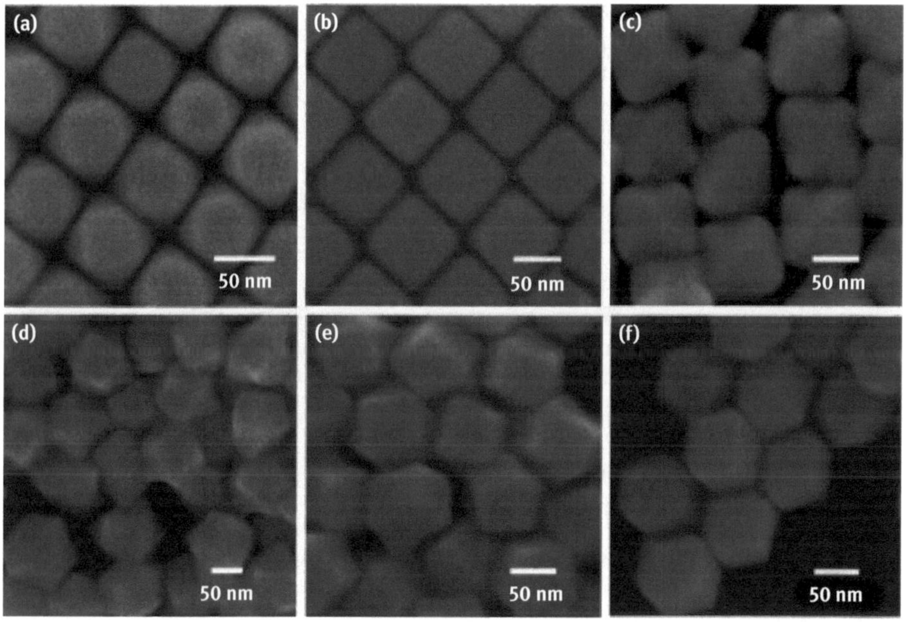

Figure 1.7: SEM images of the gold nanocrystals synthesized with shape evolution from truncated cubic to rhombic dodecahedral structures by increasing the amount of ascorbic acid added. The nanocrystals are: (a) truncated cubes, (b) cubes, (c) type I transitional product, (d) trisoctahedra, (e) type II transitional product, and (f) rhombic dodecahedra. Reproduced from H.L. Wu, C.H. Kuo and M.H. Huang, Langmuir, 2010, 26, 14, 12307. ©2010, American Chemical Society [40].

In particular, the control of the size of the nanostructures is related to the volume of the seed solution added to the growth solution. Usually, the increase in the seed concentration reflects a decrease in size. On the other hand, the geometry is tuned by a synergistic effect in the growing solution between silver ions, bromide ions, and concentration of ascorbic acid [42]. In particular, by changing the concentration of the latter (Figure 1.7), the kinetics and the thermodynamics of the reaction are well-controlled because ascorbic acid promotes the growing process on specific crystal facet of the gold seed. Moreover, the concentration of bromide ions in the growing solution (Figure 1.8) also acts as a shape-controller [43].

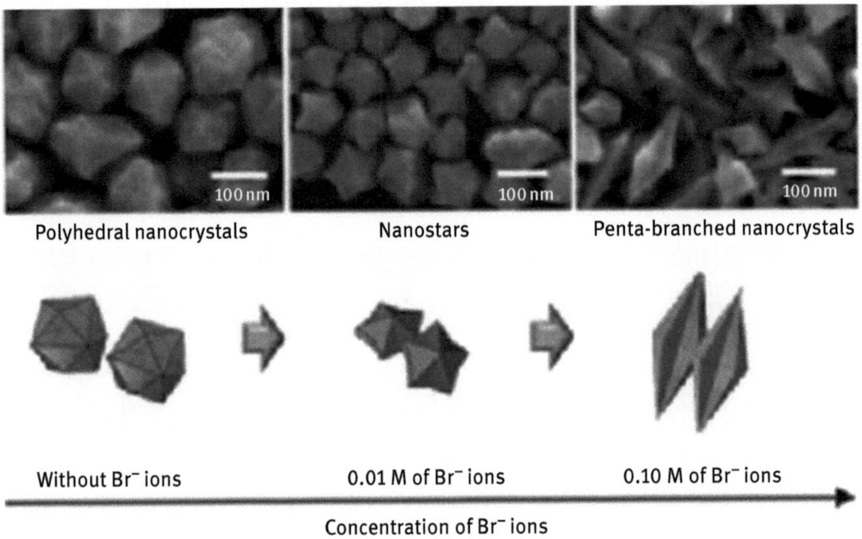

Figure 1.8: Shape evolutions from polyhedral gold nanocrystals to nanostars and penta-branched nanocrystals with increasing bromide ion concentration in the reaction solution. For the synthesis of gold nanostars, 100 µl of 0.01 M $AgNO_3$ was also added. Reproduced with permission of H-L. Wu, C-H. Chen and M.H. Huang, Chemistry of Materials, 2008, 21, 110. ©2008, American Chemical Society [43].

For this reason, Wu and colleagues have used as a capping agent CTAC in the presence of NaBr (instead of CTAB), in order to strictly control the Br^- to Ag^+ ratio [43]. On one hand, as discussed in Section 1.2.1, silver ions are critical to the promotion of the development of edges because Ag^+ adsorbs on the nanostructure's surface in the form of AgBr or AgCl, and hinders the growth of the Ag^+-passivated crystal facets [37]. On the other hand, the authors report that a better control in Br^- ions results in a fine tuning of the shape, due to the intrinsic control performed on the passivated facets [40].

Remarkably, the control over the reaction conditions and the concentration of the reactive (ascorbic acids, silver ions, seeds, CTAB, and gold precursors) may result in the production of exotic geometries, among which triangular, pod-shape, and "branched" AuNPs (Figures 1.7 and 1.8) [40, 42, 44, 45]. These particular geometries are not discussed here because the synthetic processes show very slight variations with respect to the rod- and cube-shape synthesis.

Seo and coworkers proposed another way for the synthesis of gold cubes, octahedrons, and cuboctrahedrons, which use a polyol route at high temperature in the presence of silver ions [41]. In this process, the diol is used both as a solvent and as a reducing agent, in the presence of polyvinylpyrrolidone (PVP, a surface regulating polymer). The use of the diol is also required due to its high boiling temperature (for 1,5-pentanediol, PD, 197 °C). The high temperature: (i) promotes the formation of more thermodynamically stable structures (Figure 1.9) resulting in uniform single-crystalline products, and (ii) reduces the formation of less stable twinned particles such as decahedra and rods.

The reaction is a multi-step process in which the gold precursor is reduced from Au(III) to Au(I) and eventually to Au^0, while PD is oxidized. The high temperature increases the kinetics of the gold nanocrystals formation, and the various gold intermediates generated during the reaction do not affect the final product morphology. On the other hand, PVP concentration affects the nanocrystal size, still preserving the shape. This behavior is probably due to the nonselective PVP binding to the surface of the gold seeds, which results in a reduced crystal growth along all directions.

As in the case of gold nanorods, the introduction of $AgNO_3$ may suppress the growth of a certain face of the seeds, and greatly influence the shape and surface structure of the final nanocrystals. The nanocrystal shape changed from octahedral to truncated octahedral, cuboctahedral, cubic, and spherical by increasing the amount of $AgNO_3$ (Figure 1.10) [41].

Regarding the surface structure (pointing out that {hkl} describes the orientation of the planes in a crystal by Miller indices), an octahedron has only {111} faces exposed to the surface. The {100} surface fraction continuously increases by increasing the $AgNO_3$ amount up to complete {100} coverage, that is, a cube. Overall, Ag species generated from $AgNO_3$ enhance the selective growth of {100} and/or suppress the growth of {111}. X-ray photoelectron spectroscopy analysis of the gold cubes indicates that the silver species participate in the redox reaction of the gold precursors that occurred on the seed surface. Ag^+ is readily reduced to Ag^0 by PD at high temperatures, and Ag^0 is oxidized again to Ag^+ by a galvanic exchange reaction with $AuCl_4^-$. If the silver concentration is too high and exceeds the selective deposition condition, seed growth is completely restricted along all directions, leading to small spherical particles. Furthermore, particle edge sharpening occurred concomitantly with particle growth (Figure 1.11). Overall, the total reaction is a rapid seed formation, followed by an edge sharpening, and shape/size focusing by subsequent Ostwald ripening [46].

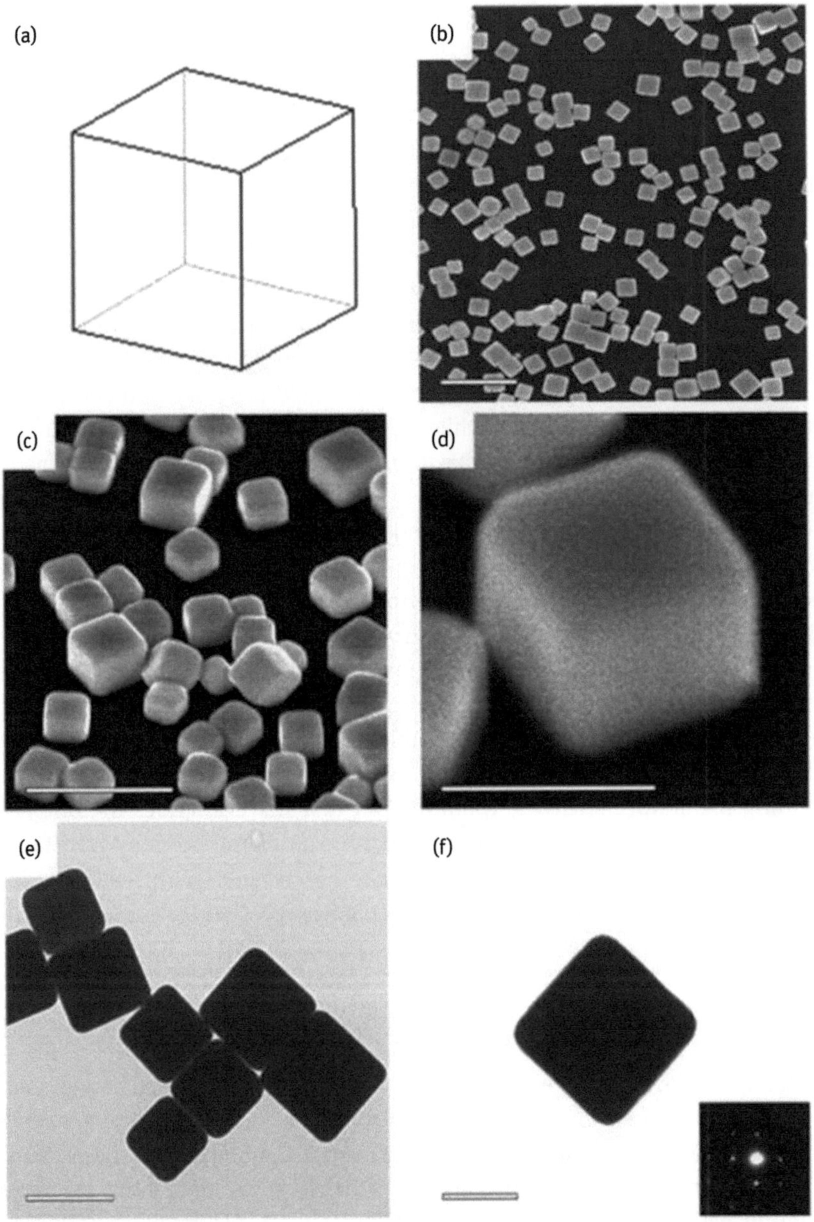

Figure 1.9: Gold nanocubes. (a) Ideal cube structure. (b, d) SEM images of the cubes. (c, d) SEM images of cubes 45° tilted. (e, f) TEM images of the cubes. Inset in (f) is a diffraction pattern. The bars indicate 500 nm (b, c), 200 nm (d, e), and 100 nm (f). Reproduced with permission of D. Seo, J.C. Park and H. Song, Journal of the American Chemical Society, 2006, 128, 14863. ©2006, American Chemical Society [41].

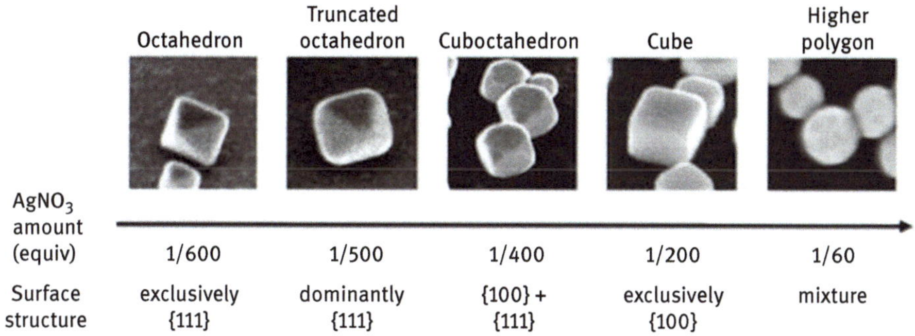

Figure 1.10: Polyhedral structures of AuNPs with respect to the amount of $AgNO_3$ added in to the reaction mixture. Reproduced with permission of D. Seo, J.C. Park and H. Song, Journal of the American Chemical Society, 2006, 128, 14863. ©2006, American Chemical Society [41].

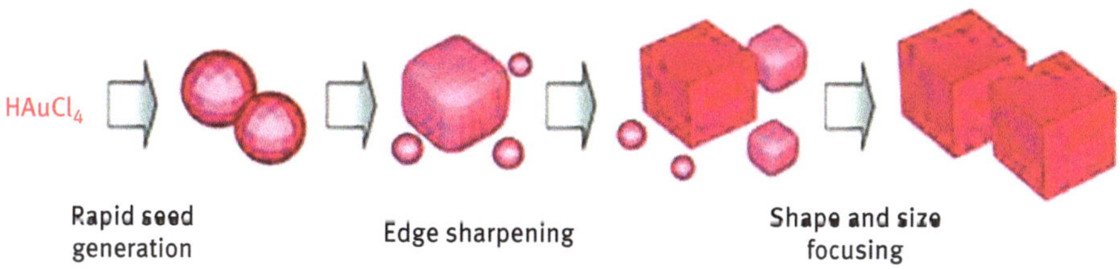

Figures 1.11: Proposed process for the polyol-route formation of gold nanocubes. Reproduced with permission of D. Seo, J.C. Park and H. Song, Journal of the American Chemical Society, 2006, 128, 14863. ©2006, American Chemical Society [41].

1.2.3 Gold nanocages

Xia and coworkers proposed a polyol synthetic route in which the preparation of gold nanocages is linked to the synthesis of high quality silver nanocubes that serve as sacrificial templates during a galvanic replacement reaction (Figure 1.12) [47].

The polyol reduction is a robust route to produce high-quality silver nanocubes in large quantities. In a typical polyol synthesis, ethylene glycol (EG) acts as both the solvent and reducing agent as in eq. (1.8) [48]:

$$2\,Ag^+ + HOCH_2CHO + H_2O \rightarrow 2\,Ag + HOCH_2COOH + 2\,H^+ \qquad (1.8)$$

where the glycolaldehyde ($HOCH_2CHO$) is formed due to the oxidation of EG by oxygen from the air. The reduction typically generates seeds with different structures: single-crystal, single-twinned, and multiple-twinned. These structures can further grow into cubes, bipyramids, and pentagonal wires, respectively [49]. In order to produce silver nanocubes in high yields, the general approaches are to: (i) remove the twinned seeds by oxidative etching, or (ii) perform a fast reduction through

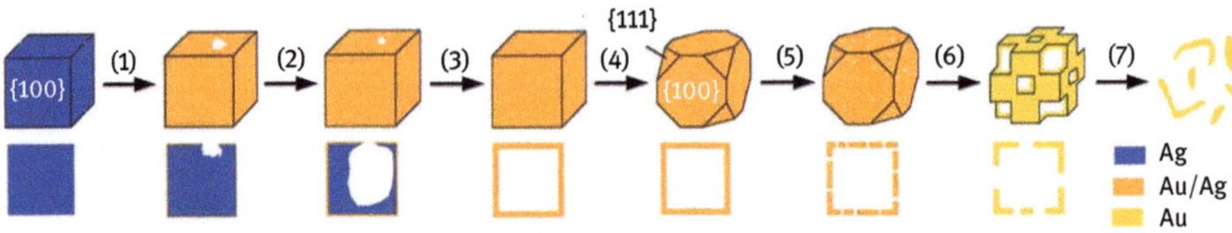

Figure 1.12: Schematic illustration summarizing how hollow nanostructures with various porosities evolve from Ag nanocubes with increasing amounts of $HAuCl_4$ solution added to the reaction. The major steps include the (1) initiation of replacement reaction at a specific site with the highest surface energy; (2) continuation of the replacement reaction between Ag and $HAuCl_4$ and the formation of a partially hollow nanostructure; (3) formation of nanoboxes with a uniform, smooth, homogeneous wall composed of Au/Ag alloy; (4) initiation of dealloying and corner reconstruction of the Au/Ag nanobox; and (5, 6) continuation of dealloying and formation of an Au nanocage–nanobox with pores in the walls. The cross-sectional views correspond to the plane along dashed lines. Note that the size of the initial Ag nanocubes will influence the amount of $HAuCl_4$ solution necessary to achieve a particular morphology. Reproduced with permission of S.E. Skrabalak, L. Au, X.D. Li and Y. Xia, Nature Protocols, 2007, 2, 2182. ©2007, Nature [48].

in situ formation or direct addition of external clusters. In the first approach, oxidative etchants (e.g., Cl^- ions combined with O_2) are introduced to selectively eliminate the twinned seeds due to the high reactivity of defects on their surface. The remaining single-crystal seeds can lead to the growth of nanocubes with the assistance of PVP, which selectively binds to the {100} facets of the silver nanostructures. The oxidative etching is typically a slow process and the synthesis usually takes more than 10 h. The second approach is based on the formation of Ag_2S clusters through the addition of Na_2S directly in the reaction solution during the nucleation of Ag atoms. Sodium sulfide significantly increases the rate of the reduction process. Thus, the rapid reduction effectively limits the formation of twinned seeds, supporting the formation of thermodynamically favored single-crystal seeds, that is, the growth of nanocubes. At the very beginning, $AgNO_3$ was used as a precursor for silver ions. On the other hand, silver nitrate resulted in low yield production of silver nanocubes, because the reaction is highly sensitive to many conditions, including salt impurities and the amount of oxygen [47]. The nitrate group is also an oxidizing agent that may cause etching of the seeds, leading to poor reproducibility. Moon and coworkers improved the reproducibility of the polyol synthesis of Ag nanocubes by employing silver trifluoroacetate (CF_3COOAg) as precursor instead of $AgNO_3$ [50]. Interestingly, the total reaction time to produce nanocubes in the range of 30–70 nm is, not comprising the preheating of EG to 150 °C and the addition of all the reactants, about 1 h.

Gold nanocages (Figure 1.13) are prepared via the galvanic replacement reaction (eq. (1.9)) between Ag cubes (as sacrificial templates) and $HAuCl_4$, which is driven by the difference in electrochemical potential between Ag/Ag^+ (0.80) and $Au/AuCl_4$- (1.00 V) [48]:

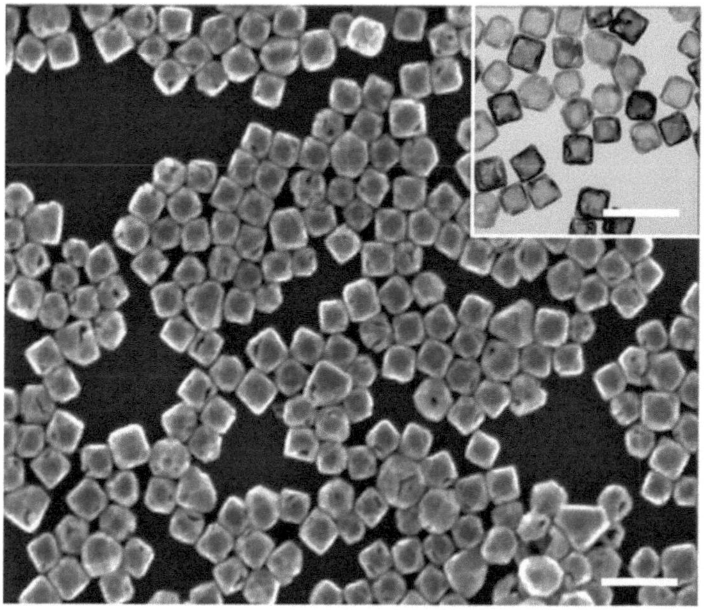

Figure 1.13: A SEM image (insert: TEM image) of Au nanocages prepared through the galvanic replacement reaction between Ag nanocubes and $HAuCl_4$ aqueous solution (scale bars are 100 nm for both images). Reproduced by permission of S.E. Skrabalak, L. Au, X. D. Li and Y. Xia, Nature Protocols, 2007, 2, 2182. ©2007, Nature [48].

$$3\ Ag_s + AuCl_4{}^- \rightarrow Au_s + 3\ Ag^+ + 4\ Cl^- \qquad (1.9)$$

To maintain epitaxial growth for the Au atoms on the Ag nanocubes, Xia and coworkers performed the reaction at 100 °C, in order to avoid any precipitation of AgCl (solubility constant at 100 °C, $K_{sp}{}^{100} = 1 \times 10^{-6}$) that is formed during the galvanic replacement. Once cooled down to room temperature, the AgCl solid is generally redissolved by addition of a saturated sodium chloride (NaCl) solution, which promotes the formation of a soluble coordination complex with chloride.

1.2.4 Gold nanoshell

Gold nanoshells are produced in two different types, accordingly to their final application: hollow or filled with other materials. For the first type, the shell is built on a sacrificial core that is subsequently removed, while the latter type has a core/shell moiety. In both cases, the core can be composed by nanoscale materials such as silver or cadmium sulfide colloids, silica beads, polymers, or metal-oxides such as magnetite [51].

In a typical procedure, the surfaces of the templates are coated with thin layers of the desired material (or its precursor) to form core-shell nanostructure (Figure 1.14).

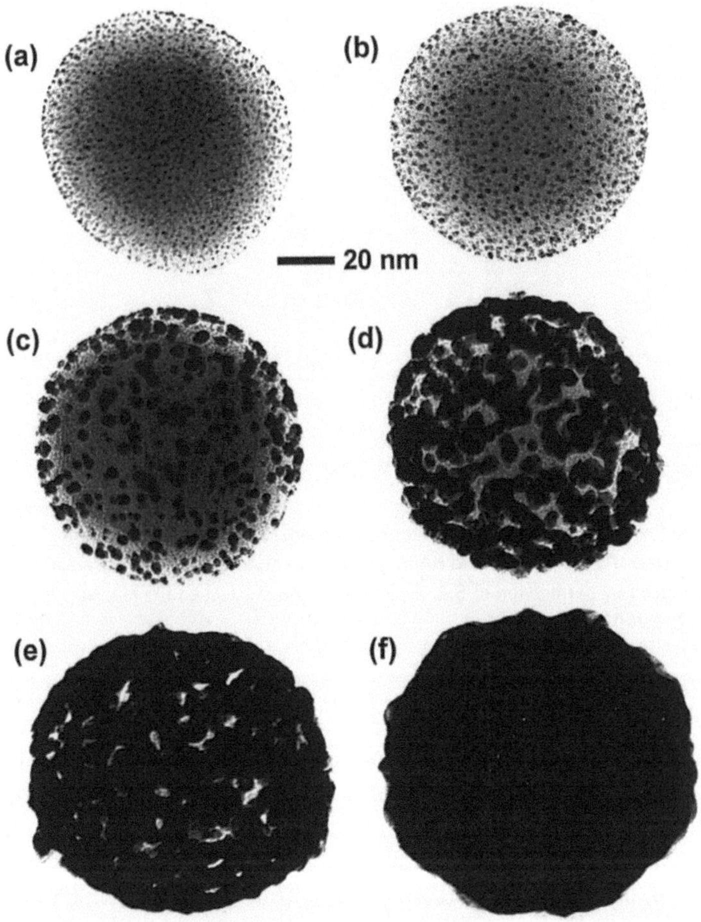

Figure 1.14: (a–f) TEM images of nanoshell growth on 120 nm diameter silica nanoparticle. (a) Initial gold colloid-decorated silica nanoparticle. (b–e) Gradual growth and coalescence of gold colloid on silica nanoparticle surface. (f) Completed growth of metallic nanoshell. Reproduced by permission of S.J. Oldenburg, R.D. Averitt, S.L. Westcott and N.J. Halas, Chemical Physics Letters, 1998, 288, 243. ©1998, Elsevier [52].

An example of this strategy was proposed by Oldenburg and coworkers [52]. They synthesized silica/gold core/shell nanoparticles. Monodisperse silica nanoparticles were produced by the Stöber method as dielectric cores. A layer of organosilane molecules (3-aminopropyltriethoxysilane, APTES) were then adsorbed on to those templates. APTES bonded to the surface of the silica nanoparticles, thanks to the silane groups, while the amine groups point outward as a new termination of the nanoparticle surface. Subsequently, modified silica nanoparticles were coated by gold clusters (1–2 nm in diameter). Gold clusters were adsorbed on the surface of silica nanoparticles due to ionic interactions in a number most likely limited by interparticle Coulomb repulsion (Figure 1.14). A subsequent sodium borohydride reduction of a gold precursor, such as chloroauric acid, in the presence of potassium carbonate resulted in the growth of a gold shell around the silica core, due to the catalytic effect of gold nucleation sites on the silica nanoparticle surface (Figure 1.14).

In order to obtain hollow nanostructures from the core/shell nanoparticles, the template-core is generally removed by calcination at high temperature in air or by selective etching in an appropriate solvent [52]. It is important to notice that the size and shape of the shell is determined by the geometry of the templates. The major problem associated with the shell growing on templates is the polycrystalline nature of the produced nanoshell. It is worth to notice that in some cases the nanoshell is made of discrete gold colloids (or domains) characterized by poor connections among themselves (Figure 1.14). Sun and coworkers avoided this problem by developing a new process based on template-engaged replacement reactions, which generate hollow nanostructures of noble metals with defined geometry and homogeneous, highly crystalline walls [53].

The key reaction involved in this process is the galvanic replacement previously discussed in Section 1.2.3 for gold nanocages (eq. (1.9)). Briefly, silver nanostructures can be oxidized by $HAuCl_4$ because the standard reduction potential of the $AuCl_4^-$/Au pair (0.99 V, versus standard hydrogen electrode (SHE)) is higher than that of the Ag^+/Ag pair (0.80 V, versus SHE). The metallic gold produced by the galvanic reaction is confined to the vicinity of the template surface, and it nucleates and grows into very small particles until it evolves into a thin shell around the silver template (Figure 1.15).

Nanoshells produced by this process usually have an incomplete structure at the initial stage. Indeed, Sun and coworkers found that both $HAuCl_4$ and AgCl can continuously diffuse across gold layer until the silver template has been completely consumed. When the reaction is refluxed at 100 °C for a certain time, the gold shells can reconstruct their walls into highly crystalline structures due to the Ostwald ripening process. At the same time, the surfaces of these hollow structures are smoothened, and any holes in the incomplete shells seem closed to form seamless shells (Figure 1.15). These gold shells have a morphology similar to that of the silver templates, and their void sizes are mainly determined by the size of the templates. On the basis of the stoichiometric relationship shown in eq. (1.9), the wall thickness of each gold nanoshell should be approximately one-tenth of the lateral dimension of the corresponding silver template. As for the synthesis of gold nanocages (Section 1.2.3), AgCl produced in this replacement reaction is completely soluble in water under experimental conditions [53].

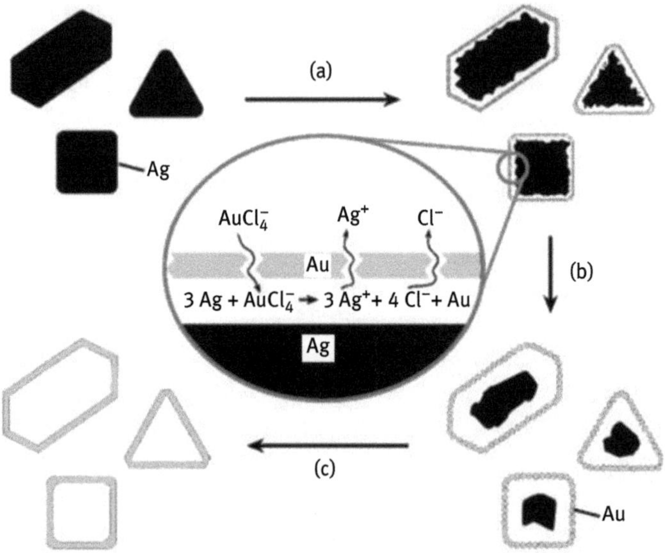

Figure 1.15: Schematic illustration of the experimental procedure that generates nanoscale shells of gold from silver templates with various morphologies. The reaction scheme is: (a) addition of $HAuCl_4$ to a dispersion of silver nanoparticles and initiation of the replacement reaction, (b) the continued replacement reaction of $HAuCl_4$ with the silver nanoparticles; (c) Depletion of silver and annealing of the resultant shells to generate smooth hollow structures. Note that the shape of each silver nanoparticle is essentially preserved in this template-engaged reaction. Reproduced by permission of Y. Sun, B.T. Mayers and Y. Xia, Nano Letters, 2002, 2, 481. ©2002, American Chemical Society [53].

It is important to notice that despite the potential of this process to produce fine-tuned and good crystallinity nanoshells, there are two important drawbacks: (i) nanostructures are only hollow, and (ii) the metal precursor must be reducible by the Ag^+/Ag pair.

1.2.5 Hybrid gold nanostructures

Very recently, disassembling noble metal nanoplatforms have attracted a growing interest due to the possibility to combine theranostics features and efficient excretion of the building blocks. A breakthrough in this topic was the introduction of the ultrasmall-in-nano approach [2]. Ultrasmall-in-nano consists in the design of biodegradable nanocapsules comprising excretable ultrasmall nanoparticles (USNPs). The key-advances of this approach are as follows: (i) USNPs mimic the light-matter interactions of bigger nanoparticles when strictly packed together [54], (ii) the size of the nanocapsules promotes accumulation in the target and cellular uptake [55], and (iii) the degradation products of the nanocapsules are quickly excretable or can be reused from organisms for physiological functions [56]. One of the most promising material developed by the ultrasmall-in-nano approach is the family of passion fruit-like nanoarchitectures (NAs) [54, 57, 58]. Their massive production, reproducibility, versatility, potential applications, biosafety, and biokinetics were already demonstrated [56, 58–63].

Gold USNPs with a size of 2.8 ± 0.4 nm are usually synthesized in aqueous environment in the presence of poly(sodium 4-styrene sulfonate) (PSS) and assembled in controlled aggregates through ionic interactions with cationic poly (L-lysine) (PL) [54]. The aggregates are then employed as templates in order to grow silica shells of about 20 nm in thickness by employing a modified Stöber method, obtaining NAs of about 100 nm in diameter. It is useful to remember here the mechanism of the standard Stöber synthesis [64]. The introduction of tetraethoxysilane (TEOS) into the reaction medium is followed by its hydrolysis reactions, resulting in orthosilicic acid. The polymerization of orthosilicic acid occurs when the concentration exceeds the saturation limit in ethanol (about 0.02–0.03%) [65]. The process yields, in sequence, low- to high-molecular polymers and, due to their condensation, particles of 1–2 nm in diameter. Then nuclei increase in size following a LaMer growth pattern until their diameter reaches a critical value of 5–7 nm, which results in their aggregation to form solid silica nanoparticles [65]. This process continues until the concentration of orthosilicic acid in the reaction medium exceeds the saturation limit. A similar mechanism also happens in water solutions, but usually with longer timeframe [66]. In the case of NAs, the composition of the silica shell around gold arrays is related to the simultaneous presence of both amines from PL and aromatic moieties from PSS [58]. Indeed, aromatic additives have been demonstrated to be required for the composition of hollow silica microspheres [67]. TEOS is hydrolyzed to orthosilicic acid and partially adsorbed into gold USNPs polymeric arrays during its polymerization. Then, silica nuclei of 1–2 nm readily formed in solution, aggregate on the external surface of the arrays and the remaining orthosilicic acid continues to form a complete shell until its saturation limit is reached or the reaction interrupted. Interestingly, the silica shell of the NAs does not show gold USNPs inclusion. NAs produced by the standard synthetic approach usually show a diameter of 107.6 ± 16.1 nm, a wall thickness of 18.9 ± 2.2 nm, and a gold content of 5.9 ± 1.3% w/w.

1.3 Methods

All the following synthetic processes were personally tested by the author.

1.3.1 Nanoparticle characterization

The basic characterizations for geometry and stability of AuNPs are UV-Vis spectra, electron microscopy, dynamic light scattering (DLS), and electrophoresis [68]. The extinction spectra of colloids are usually collected using a standard UV-Vis double-beam spectrophotometer. Spectra have important contributions from both absorption and scattering of light with respect to the ones collected from solutions of organic molecules, where the absorbance dominates [69]. The spectral features are linked to the size of the structures and to the peculiar interactions between light and nanomaterials (see Chapter 2). In general, the principal band in AuNPs spectra (extinction band) is the sum of absorption and scattering, due to the stimulation of the LSPR, while the rising of background at short wavelength results from Rayleigh scattering, an elastic phenomenon whose probability increases with the frequency of light. These two effects produce the final shape of the typical spectrum of colloids. The shape, width, and intensity of the plasmon band can give preliminary information over the shape and dispersion of AuNPs solutions and, if the molar extinction coefficient is known, on their concentration [70].

Electron microscopies, such as scanning electron microscopy (SEM) or transmission electron microscopy (TEM) are employed to precisely determine the shape and the size of nanoparticles. These techniques are usually quite expensive, relatively time-consuming, and require immobilization of the samples on suitable supports. For SEM analysis, a drop of the colloid has to be dried on a silicon chip to be imaged. For TEM analysis, the sample has to be dripped on a suitable TEM grid and imaged, for aqueous samples, after at least half an hour. Lots of software, such as the freeware ImageJ, can be employed to analyze images and evaluate, usually on 300–500 nanoparticles, the average size and the sphericity (if required) of the nanostructures.

DLS measurements provide nanoparticles hydrodynamic diameter and zeta (ζ) potential. In both cases, the light of a laser scattered from the particles in solution is measured. For particle size measurements, the scattered light is usually collected at 13° or 173° and the intensity fluctuations due to Brownian motion are determined. Brownian motion is the stochastic movement of particles caused by random collisions of the particles in the media. Smaller particles move faster than bigger ones and the relationship between the size of a particle and its diffusivity due to Brownian motion is governed by the Stokes–Einstein equation (see Chapter 2). As the particles move, the constructive and destructive interference of the scattered light will cause fluctuations in the scattered intensity. By the analysis of the time correlation in intensity fluctuations, it is possible to obtain the size distribution of AuNPs in the colloid.

The ζ-potential is a measure of the charge of nanoparticles. It can be considered the potential difference between the dispersion medium and the external part of the stationary layer of fluid complexed around the nanoparticle. In particular, this value gives an indication of the potential stability of the colloidal system (normally colloids with values more positive than +30 mV or more negative than −30 mV are considered stable) and it is usually dependent on the pH of the solution. The ζ-potential is calculated by determining the electrophoretic mobility and, then, applying the Henry equation. The electrophoretic mobility is obtained by performing an electrophoresis experiment on the liquid sample in a standard capillary cuvette and measuring the velocity of the particles using a laser Doppler velocimetry (LDV). Generally, an LDV collects the light scattered at 13° and combines it with the reference beam. This produces a fluctuating intensity signal where the rate of fluctuation is proportional to the speed of the particles and, through the Henry equation, to the ζ-potential value. Thus, DLS measurements provide indication about superficial changes on nanoparticles since a substitution in the coating or a reaction on the metal surface change the total charge and/or the hydrodynamic radius of the nanomaterial. Usually, DLS measurements are performed on 1 ml of colloidal solutions buffered at specific pH and salt concentrations.

Gel-electrophoresis is an inexpensive and fast technique to measure the size/charge ratio and the stability of nanoparticles [71, 72]. When an electric field is applied across an agarose gel, charged particles are attracted toward the electrode of the opposite charge, and reach a limit drift velocity proportional to the electric field and the nanoparticle motility because of the friction of the gel mesh. Thus, particles with the same size and charge move with the same constant velocity, and the colloid, if stable, will be separated in different bands. Indeed, if the sample is not stable, nanoparticles aggregate and may run in a smear. If samples run in bands, the difference in their retention time can be determined because it is inversely proportional to the ratio of the charge and hydrodynamic radius (inclusive of the solvation sphere). An increase of the radius usually results from volume increases due to a coating substitution or to the conjugation of the nanoparticle to (bio)molecules (dyes or proteins). In general, the larger the molecule the larger and slower will be the nanosystem. Concerning charge variation, the total charge of functionalized nanostructures at a given pH depends on the protonation state of the superficial groups and of the conjugates. Finally, electrophoretic analysis is usually performed on 0.6–2% agarose gels in TBE (trizma/borate/ethylenediaminetetraacetic acid) buffer 0.5× and applying at the electrodes a potential difference of 90 V for 30–90 min, depending on the size and geometry of the nanoparticles.

1.3.2 Turkevich method

In the standard procedure to obtain a water colloid of AuNPs with an average diameter of 20 nm, a solution of trisodium citrate (50 ml, 2.2 mM) was heated to boiling point and a solution of $HAuCl_4$ (1 ml, 25 mM) was added rapidly [7]. In about 25 s the boiling solution turns from a light yellow to faintly blue. After an additional

70 s (approximately) the blue color suddenly changes into a brilliant red, indicating the formation of spherical AuNPs. The solution was further refluxed for 15 min with stirring to promote the monodispersion of the samples.

1.3.3 Brust method

An aqueous solution of hydrogen tetrachloroaurate (30 ml, 30 mM) was mixed with a solution of tetraoctylammonium bromide in toluene (80 ml, 50 mM) [25]. The two-phase mixture was vigorously stirred until all the tetrachloroaurate was transferred into the organic layer and dodecanethiol (170 mg) was then added to the organic phase. A freshly prepared aqueous solution of sodium borohydride (25 ml, 0.4 M) was slowly added with vigorous stirring. After further stirring for 3 h the organic phase was separated, evaporated to 10 ml in a rotary evaporator and mixed with 400 ml ethanol to remove excess thiol. The mixture was kept for 4 h at −18 °C and the dark brown precipitate was filtered off and washed with ethanol. The crude product was dissolved in 10 ml of toluene and again precipitated with 400 ml ethanol, obtaining a black solid of AuNSs with an average diameter of 10 nm.

1.3.4 Xia method

An $HAuCl_4$ aqueous solution (1 ml, 0.5 wt%) and an aqueous solution of $AgNO_3$ (42.5 μl, 0.1 wt%) were added to an aqueous solution of citrate (1.5 ml, 1% wt) with stirring [10]. Note that adding the citrate solution to the $HAuCl_4/AgNO_3$ mixture solution may cause its color to change from light yellow to orange, leading to black. This mixture was incubated for 5 min before quickly adding it to a solution of citrate (50 ml, 1% wt) at 100 °C with vigorous stirring. The color of the reaction solution changed quickly from colorless, to greyish blue, to purple, and finally to ruby red within less than 1 min at a citrate concentration ranging from 0.297 to 0.00610 wt%. The reaction solution was further refluxed for 1 h under stirring to form uniform quasi-spherical 15 nm diameter AuNSs.

1.3.5 Zhong method

Gold seeds were prepared by refluxing for 30 min an aqueous solution of $HAuCl_4$ (100 ml, 1 mM) with sodium acrylate (24 mM) [20]. The reaction produced a deep-red solution (typical color for spheres with diameter from 10 to 20 nm). For the second step all the seed solution was mixed in a solution of $HAuCl_4$ (15 ml, 25 mM) and diluted in MilliQ water (845 ml). Acrylic acid solution (40 ml, 0.5 M) was added, and the mixture was adjusted to pH 7. After 3 days at 22 °C with continuous stirring, the reaction produced a wine-red solution of 30 ± 1.7 nm AuNSs at a 0.5 nM (2.92×10^{11} NPs/ml) concentration.

1.3.6 Gold nanorods

Au nanorods were synthesized using the silver ion-assisted seed-mediated method [73]. Typically, the seed solution was prepared by the addition of $HAuCl_4$ (0.25 ml, 0.01 M) to CTAB solution (10 ml, 0.1 M) in a 15 ml plastic tube with gentle mixing (CTAB solution must be pre-prepared at 60 °C). A freshly prepared, ice-cold $NaBH_4$ solution (0.6 ml, 0.01 M) was then quickly injected into the mixture, followed by rapid inversion for 2 min. The seed solution was kept at room temperature for at least 2 h before use. To grow Au nanorods, $HAuCl_4$ (2.0 ml, 0.01 M) and $AgNO_3$ (0.4 ml, 0.01 M) were mixed with CTAB (40 ml, 0.1 M) in a 50 ml plastic tube. Hydrochloric acid (0.8 ml, 1.0 M) was then added to adjust the pH of the solution to 1–2, followed by the addition of ascorbic acid (0.32 ml, 0.1 M). Finally, the seed solution (0.096 ml) was injected into the growth solution. The solution was gently mixed for 10 s and left undisturbed at room temperature for at least 6 h before its pH was adjusted to 3–4 by the addition of NaOH (0.8 ml, 1.0 M) for the storage of the final colloid.

1.3.7 Gold nanocubes

Gold nanocubes were grown following a two-step method [39]. In the first step, gold seeds were prepared by the addition of a freshly prepared, ice-cold aqueous $NaBH_4$ solution (0.6 ml, 0.01 M) into an aqueous mixture (10 ml) composed of $HAuCl_4$ (0.01 M) and CTAB (0.1 M), followed by rapid mixing for 2 min (CTAB solution must be pre-prepared at 60 °C). The resultant brownish seed solution was kept at room temperature for 1 h before use in order to decompose the excess borohydride. The growth solution was prepared by the sequential addition of CTAB (6.4 ml, 0.1 M), $HAuCl_4$ (0.8 ml, 0.01 M), and ascorbic acid (3.8 ml, 0.1 M) into water (32 ml). The CTAB-stabilized seed solution was diluted 10 times with water. The diluted seed solution was then added (in a range from 0.02 to 0.8 ml) into the growth solution. The resultant mixture was mixed by gentle inversion for 10 s and then left undisturbed overnight. The average edge length of the Au nanocubes produced was from 45 to 70 nm.

1.3.8 Gold nanocubes and polyhedrons

The gold seed solution was prepared with the same procedure described in Section 1.3.6 for the synthesis of gold nanorods, while the growing process is explained in Wu and coworkers [40]. Note that the final morphology (cubes, bipyramids, truncated cubes, stars, and rhombic dodecahedral) is decided by the amount of ascorbic acid added to the growing solution [40, 43]. Two vials were labeled A and B, and a growth solution was prepared in each of the two vials. First, CTAC surfactant (0.32 g, final concentration 0.10 M) was added. Depending on the morphology of gold nanocrystals to be synthesized, slightly different volumes of deionized water (9.640–9.565 ml) were added to each vial. The vials were then kept in a water bath at 30 °C. To both vials were added $HAuCl_4$ solution (250 µl, 0.01 M), $AgNO_3$ (100 µl, 0.01 M), and sodium bromide (NaBr, 10 µl, 0.01). Finally, ascorbic acid (75–150 µl, 0.04 M) was introduced. For example, for the synthesis of gold nanocubes, 90 µl of ascorbic acid was used, whereas 150 µl of ascorbic acid was added for the growth of rhombic dodecahedra. The total solution volume in each vial was 10 ml. The color of the solution turned colorless after the addition of ascorbic acid, indicating the reduction of Au^{3+} to Au^{+} species. Next, 25 µl of the seed solution was added to the solution in vial A with shaking until the solution color turned light pink (~5 s). Then 25 µl of the solution in vial A was transferred to vial B with thorough mixing for ~10 s. The solution in vial B was left undisturbed for 15 min for particle growth and the resulting nanostructures purified by 10 min cycles of centrifugation at 3,000 rpm. The gold nanocubes and rhombic dodecahedra have average sizes of 72 and 74 nm, respectively. To make smaller gold nanocubes and rhombic dodecahedra, the volume of the seed solution used needs to be varied (e.g., for the preparation of gold nanocubes with average sizes of 40 nm, 65 µl of the seed solution were added to the growth solution in vial A).

1.3.9 Gold nanocages

The reaction procedure is a two-step method in which the first reaction serves to produce the silver nanocubes (the sacrificial templates), and in the second gold nanocages are produced through the galvanic replacement reaction [48].

A solution of EG (6 ml) was warmed at 150 °C for 1 h in order to allow the water vapor to escape. Then, 80 µl of a 3 mM EG solution of Na_2S were added maintaining the heating. After 10 min the following were added to the solution: (i) 1.5 ml of a 0.67 mM EG solution of PVP 30k, and (ii) 0.5 ml of 0.28 M of EG solution of silver nitrate. The solution turned to a green-ochre color in 15–20 min and the reaction was quenched by stopping the heating, in order to reach room temperature. Acetone (15 ml) was added and the solution centrifugated (2000 × g for 30 min). The precipitate was dispersed in milliQ water and purified by three cycles of precipitation/redispersion in milliQ water by centrifugation (10 min at 9,000 rpm). At the final step the precipitate was dissolved in 4 ml of milliQ water.

An aqueous solution of PVP (5 ml, 9 mM) was mixed with 100 µl of the as-synthesized silver nanocubes colloid, and heated to 100 °C. After 10 min, an aqueous solution of gold tetrachloroaurate (0.1 M) was added at a constant rate of

0.75 ml/min. By adding the gold salt, the solution varies its color from ocher to yellow/blue. Remember that the gold precursor volume could vary from 3 to 8 ml according to the type of cages desired. After further 10 min at 100 °C, the solution was left to reach room temperature, and then NaCl was added till saturation. In order to purify the structures, the solution was centrifuged at 2,000 rpm for 30 min. The precipitate (a mix of gold nanocages and NaCl) was dispersed in water, and purified by five cycles of centrifugation/redispersion in water.

1.3.10 Passion fruit-like nanoarchitectures (NAs)

The following protocol is standardized for the production of 1.5 mg NAs in about 4 h. The protocol can be scaled-up to 20 mg [54, 61, 62].

Synthesis of AuNPs: Ultrasmall AuNPs with a diameter of approximately 3 nm were prepared according to the following procedure. To 20 mL of milliQ water were added 10 µL of poly(sodium 4-styrene sulfonate) (70 kDa, 30% aqueous solution, PSS) and 200 µL of $HAuCl_4$ aqueous solution (10 mg/mL). During vigorously stirring, 200 µL of sodium borohydride (8 mg/mL in milliQ water) was added quickly, and the mixture was vigorously stirred for 2 min. After the addition of $NaBH_4$, the solution underwent some color changes until becoming brilliant orange. Before its use the solution was aged for 10 min and employed without further purification.

Synthesis of AuNPs arrays: About 20 mL of AuNPs solution was added to a 50 mL round-bottomed flask, followed by 200 µL water solution of poly(L-lysine) hydrobromide 15–30 kDa (PL, 20 mg/mL), and the mixture was allowed to stir for 20 min at room temperature. The as-synthesized gold aggregates were collected by centrifugation (13,400 rpm for 3 min), suspended in 2 mL of milliQ water and sonicated for maximum 4 min.

Synthesis of NAs: About 70 mL of absolute ethanol followed by 2.4 mL of ammonium hydroxide solution (30% in water), and 40 µL of tetraethyl orthosilicate (TEOS, 98%) were added in two 50 mL plastic Falcon tubes. The solution was allowed to stir for 5 min at RT. About 2 mL of the AuNPs arrays previously prepared were added to the Falcon (1 mL each) and the solution was allowed to gently shake for further 3 h. The as-synthesized NAs were collected by 30 min centrifugation at 4,000 rpm, washed twice with ethanol to remove unreacted precursors and suspended in 1 mL of ethanol. A short spin centrifugation was employed to separate the structures over 150 nm from the supernatant, which was recovered as a pink-iridescent solution. The solution containing about 1.5 mg of NAs was stored at –20 °C until the employment. It remains usually stable for at least 1 year. Product recovery: (i) 2 min centrifugation at 13400 rpm, (ii) remove the colorless supernatant, and (iii) add the solvent of interest. The solubility of NAs in water, buffers, and physiological fluids is tested for up to 60 mg/mL.

References

[1] Jain PK, Lee KS, El-Sayed IH, El-Sayed MA. Calculated absorption and scattering properties of gold nanoparticles of different size, shape, and composition: applications in biological imaging and biomedicine. J Phys Chem B 2006;110(14):7238–7248.
[2] Cassano D, Pocoví-Martínez S, Voliani V. Ultrasmall-in-nano approach: enabling the translation of metal nanomaterials to clinics. Bioconjug Chem 2018;29(1):4–16.
[3] Vlamidis Y, Voliani V. Bringing again noble metal nanoparticles to the forefront of cancer therapy. Front Bioeng Biotechnol 2018;6:143.
[4] Yin Y, Alivisatos AP. Colloidal nanocrystal synthesis and the organic-inorganic interface. Nature 2005;437(7059):664–670.
[5] Boisselier E, Astruc D. Gold nanoparticles in nanomedicine: preparations, imaging, diagnostics, therapies and toxicity. Chem Soc Rev 2009;38(6):1759–1782.
[6] Kim D, Kim K. Controlled synthesis and biomolecular probe application of gold nanoparticles. Micron 2011;42(3):207–227.
[7] Enustun B, Turkevich J. Coagulation of colloidal gold. J Am Chem 1963;85(21):3317–3328.
[8] Ji X, Song X, Li J, Bai Y, Yang W, Peng X. Size control of gold nanocrystals in citrate reduction: the third role of citrate. J Am Chem Soc 2007;129(45):13939–13948.
[9] Goia DV, Matijevic E. Tailoring the particle size of monodispersed colloidal gold. Colloids Surfaces A Physicochem Eng Asp 1999;146(1):139–152.
[10] Xia HB, Bai SO, Hartmann J, Wang DY. Synthesis of monodisperse quasi-spherical gold nanoparticles in water via silver (I)-assisted citrate reduction. Langmuir 2010;26(5): 3585–3589.
[11] Nel AE, Mädler L, Velegol D et al. Understanding biophysicochemical interactions at the nano–bio interface. Nat Mater 2009;8(7):543–557.
[12] LaMer V, Dinegar R. Theory, production and mechanism of formation of monodispersed hydrosols. J Am Chem 1950;72(11):4847–4854.
[13] Privman V, Goia D, Park J, Matijević E. Mechanism of formation of monodispersed colloids by aggregation of nanosize precursors. J Colloid Interface Sci 1999;213(1):36–45.
[14] Wiig E. Carbon dioxide cleavage from acetone dicarboxylic acid. J Phys Chem 1928;32(7): 961–981.
[15] Wu X, Redmond PL, Liu H, Chen Y, Steigerwald M, Brus L. Photovoltage mechanism for room light conversion of citrate stabilized silver nanocrystal seeds to large nanoprisms. J Am Chem Soc 2008;130(29):9500–9506.
[16] Sanchez C, Del PM, Leiva E. An embedded atom approach to underpotential deposition phenomena. Surf Sci 1999;421(1–2):59–72.
[17] Perrault SD, Chan WCW. Synthesis and surface modification of highly monodispersed, spherical gold nanoparticles of 50–200 nm. J Am Chem Soc 2009;131(47):17042–17043.
[18] Li CC, Shuford KL, Chen MH, Lee EJ, Cho SO. A facile polyol route to uniform gold octahedra with tailorable size and their optical properties. ACS Nano 2008;2(9):1760–1769.
[19] Frens G. Controlled nucleation for the regulation of the particle size in monodisperse gold suspensions. Nature 1973;241(105):20–22.
[20] Njoki P, Lim I, Mott D, Park H. Size correlation of optical and spectroscopic properties for gold nanoparticles. J Phys Chem C 2007;(111):14664–14669.
[21] Takacs J, Xt J. The Wacker reaction and related alkene oxidation reactions. Curr Org Chem 2003;7:369.
[22] Njoki PN, Luo J, Kamundi MM, Lim S, Zhong CJ. Aggregative growth in the size-controlled growth of monodispersed gold nanoparticles. Langmuir 2010;26(16):13622–13629.

[23] Matacotta FC, Ottaviani G. Science and Technology of Thin Films, River Edge, NJ, World Scientific, 1995.
[24] Sau T, Pal A, Jana N, Wang Z, Pal T. Size controlled synthesis of gold nanoparticles using photochemically prepared seed particles. J Nanoparticle ... 2001;3:257.
[25] Brust M, Walker M, Bethell D, Schiffrin DJ, Whyman R. Synthesis of thiol-derivatised gold nanoparticles in a two-phase Liquid–Liquid system. J Chem Soc, Chem Commun 1994;(7): 801–802.
[26] Pellegrino T, Manna L, Kudera S et al. Hydrophobic nanocrystals coated with an amphiphilic polymer shell: A general route to water soluble nanocrystals. Nano Lett 2004;4(4):703–707.
[27] Yu YY, Chang SS, Lee CL, Wang CRC. Gold nanorods: electrochemical synthesis and optical properties. J Phys Chem B;1997.
[28] Pérez-Juste J, Pastoriza-Santos I, Liz-Marzán LM et al. Gold nanorods: Synthesis, characterization and applications. Coord Chem Rev 2005;249(17–18):1870–1901.
[29] Törnblom M, Henriksson U. Effect of solubilization of aliphatic hydrocarbons on size and shape of rodlike C 16 TABr micelles studied by 2 H NMR relaxation. J Phys Chem B 1997;101 (31):6028–6035.
[30] Henglein A, Giersig M. Formation of colloidal silver nanoparticles: capping action of citrate. J Phys Chem B 1999;103(44):9533–9539.
[31] Durr NJ, Larson T, Smith DK, Korgel B a, Sokolov K, Ben-Yakar A. Two-photon luminescence imaging of cancer cells using molecularly targeted gold nanorods. Nano Lett 2007;7(4): 941–945.
[32] Scarabelli L, Sánchez-Iglesias A, Pérez-Juste J, Liz-Marzán LM. A “Tips and Tricks” practical guide to the synthesis of gold nanorods. J Phys Chem Lett 2015;6(21):4270–4279.
[33] Nikoobakht B, El-Sayed M a. Evidence for bilayer assembly of cationic surfactants on the surface of gold nanorods. Langmuir 2001;17(20):6368–6374.
[34] Johnson C, Dujardin E. Growth and form of gold nanorods prepared by seed-mediated, surfactant-directed synthesis. J Mater Chem 2002;12:6.
[35] Gao J, Bender CM, Murphy CJ. Dependence of the gold nanorod aspect ratio on the nature of the directing surfactant in aqueous solution. Langmuir 2003;19(21):9065–9070.
[36] Torigoe K, Esumi K. Preparation of colloidal gold by photoreduction of tetracyanoaurate(1-)-cationic surfactant complexes. Langmuir 1992;8(1):59–63.
[37] Jana NR, Gearheart L, Murphy CJ. Seed-mediated growth approach for shape-controlled synthesis of spheroidal and rod-like gold nanoparticles using a surfactant template. Adv Mater 2001;13(18):1389–1393.
[38] Pal T, Jana NR, et al. Organized media as redox catalysts. Langmuir 1998;14(17):4724–4730.
[39] Wu X, Ming T, Wang X, Wang P, Wang J, Chen J. High-photoluminescence-yield gold nanocubes: for cell imaging and photothermal therapy. ACS Nano 2010;4(1):113–120.
[40] Wu H-L, Kuo C-H, Huang MH. Seed-mediated synthesis of gold nanocrystals with systematic shape evolution from cubic to trisoctahedral and rhombic dodecahedral structures. Langmuir 2010;26(14):12307–12313.
[41] Seo D, Park JC, Song H. Polyhedral gold nanocrystals with O h symmetry: from octahedra to cubes. J Am Chem Soc 2006;128(46):14863–14870.
[42] Sau TK, Murphy CJ. Room temperature, high-yield synthesis of multiple shapes of gold nanoparticles in aqueous solution. J Am Chem Soc 2004;126(28):8648–8649.
[43] Wu H-L, Chen C, Huang MH. Seed-mediated synthesis of branched gold nanocrystals derived from the side growth of pentagonal bipyramids and the formation of gold nanostars. Chem Mater 2009;21(1):110–114.
[44] Hao E, Bailey RC, Schatz GC, Hupp JT, Li S. Synthesis and optical properties of “Branched” gold nanocrystals. Nano Lett 2004;4(2):327–330.

[45] Chen S, Wang ZL, Ballato J, Foulger SH, Carroll DL. Monopod, bipod, tripod, and tetrapod gold nanocrystals. J Am Chem Soc 2003;125(52):16186–16187.
[46] Hoang TKN, Deriemaeker L, La VB, Finsy R. Monitoring the simultaneous ostwald ripening and solubilization of emulsions. Langmuir 2004;20(21):8966–8969.
[47] Sun Y, Mayers B, Xia Y. Transformation of silver nanospheres into nanobelts and triangular nanoplates through a thermal process. Nano Lett 2003;3(5):675–679.
[48] Skrabalak SE, Au L, Li XD, Xia Y. Facile synthesis of Ag nanocubes and Au nanocages. Nat Protoc 2007;2(9):2182–2190.
[49] Chen J, Yang M, Zhang Q et al. Gold nanocages: a novel class of multifunctional nanomaterials for theranostic applications. Adv Funct Mater 2010;20(21):3684–3694.
[50] Moon GD, Choi S-W, Cai X et al. A new theranostic system based on gold nanocages and phase-change materials with unique features for photoacoustic imaging and controlled release. J Am Chem Soc 2011;133(13):4762–4765.
[51] Bardhan R, Chen W, Bartels M et al. Tracking of multimodal therapeutic nanocomplexes targeting breast cancer in vivo. Nano Lett 2010;10(12):4920–4928.
[52] Oldenburg S, Averitt R. Nanoengineering of optical resonances. Chem Phys Lett 1998;288 (2–4):243–247.
[53] Sun Y, Mayers BT, Xia Y. Template-engaged replacement reaction: a one-step approach to the large-scale synthesis of metal nanostructures with hollow interiors. Nano Lett 2002;2(5): 481–485.
[54] Cassano D, Rota Martir D, Signore G, Piazza V, Voliani V. Biodegradable hollow silica nanospheres containing gold nanoparticle arrays. Chem Commun 2015;51(49):9939–9941.
[55] Mapanao AK, Santi M, Faraci P, Cappello V, Cassano D, Voliani V. Endogenously triggerable ultrasmall-in-nano architectures: targeting assessment on 3D pancreatic carcinoma spheroids. ACS Omega 2018;3(9):11796–11801.
[56] Cassano D, Summa M, Pocoví-Martínez S et al. Biodegradable ultrasmall-in-nano gold architectures: mid-period in vivo distribution and excretion assessment. Part Part Syst Charact 2019;36(2):1800464.
[57] Cassano D, Santi M, D'Autilia F, Mapanao AK, Luin S, Voliani V. Photothermal effect by NIR-responsive excretable ultrasmall -in-nano ultrasmall-in-nano" architectures. Mater Horizons 2019;6(3):531–537.
[58] Cassano D, Santi M, Cappello V, Luin S, Signore G, Voliani V. Biodegradable passion fruit-like nano-architectures as carriers for cisplatin prodrug. Part Part Syst Charact 2016;33(11): 818–824.
[59] Avigo C, Cassano D, Kusmic C, Voliani V, Menichetti L. Enhanced photoacoustic signal of passion fruit-like nanoarchitectures in a biological environment. J Phys Chem C 2017;121(12): 6955–6961.
[60] d'Amora M, Cassano D, Pocoví-Martínez S, Giordani S, Voliani V. Biodistribution and biocompatibility of passion fruit-like nano-architectures in zebrafish. Nanotoxicology 2018:1–9.
[61] Pocoví-Martínez S, Cassano D, Voliani V. Naked nanoparticles in silica nanocapsules: a versatile family of nanorattle catalysts. ACS Appl Nano Mater 2018;1(4):1836–1840.
[62] Cassano D, David J, Luin S, Voliani V. Passion fruit-like nano-architectures: a general synthesis route. Sci Rep 2017;7:43795.
[63] Cassano D, Rota Martir D, Signore G et al. Biodegradable nano-architectures containing gold nanoparticles arrays. MRS Adv 2016;1(30):2173–2179.
[64] Stöber W, Fink A, Bohn E. Controlled growth of monodisperse silica spheres in the micron size range. J Colloid Interface Sci 1968;26(1):62–69.
[65] Masalov VM, Sukhinina NS Kudrenko E a, Emelchenko G a. Mechanism of formation and nanostructure of Stöber silica particles. Nanotechnology 2011;22(27):275718.

[66] Begarani F, Cassano D, Margheritis E, Marotta R, Cardarelli F, Voliani V. Silica-based nanoparticles for protein encapsulation and delivery. Nanomaterials 2018;8(11):886.
[67] Bommel KJC, van, Jung JH, Shinkai S. Poly(L-lysine) aggregates as templates for the formation of hollow silica spheres. Adv Mater 2001;13(19):1472–1476.
[68] Faria M, Björnmalm M, Thurecht KJ et al. Minimum information reporting in bio–nano experimental literature. Nat Nanotechnol 2018;13(9):777–785.
[69] Eustis S, el-Sayed MA. Why gold nanoparticles are more precious than pretty gold: noble metal surface plasmon resonance and its enhancement of the radiative and nonradiative properties of nanocrystals of different shapes. Chem Soc Rev 2006;35(3):209–217.
[70] Link S, El-Sayed M. Spectral properties and relaxation dynamics of surface plasmon electronic oscillations in gold and silver nanodots and nanorods. J Phys Chem B 1999;(1): 8410–8426.
[71] Voliani V, Luin S, Ricci F, Beltram F. Single-step bifunctional coating for selectively conjugable nanoparticles. Nanoscale 2010;2(12):2783–2789.
[72] Voliani V, Ricci F, Luin S, Beltram F. Peptidic coating for gold nanospheres multifunctionalizable with photostable and photolabile moieties. J Mater Chem 2012;22 (29):14487.
[73] Ni W, Kou X, Yang Z, Wang J. Tailoring longitudinal surface plasmon wavelengths, scattering and absorption cross sections of gold nanorods. ACS Nano 2008;2(4):677–686.

2 Behaviors of gold nanoparticles

The behavior of the matter at the nanoscale is often unexpected and different from that of bulks. In any case, nanomaterials have been investigated for the production of novel materials with peculiar features. The properties of metal colloids, such as localized surface plasmon resonance (LSPR) modes and catalytic activity, depend on their size and shape. Indeed, as discussed in Chapter 1, great efforts have been devoted to the development of methods for the production of controlled nanomaterials and for their surface modification (Section 2.2). It is worth to remember that a tight control over the production processes is necessary to achieve the desired colloids, and, thus, to finely tune their physical–chemical properties [1, 2].

In this chapter, the photophysical properties of gold nanostructures, with a special focus on the correlation between geometries and optical features, are introduced in Section 2.1. In the first part of this section, a general description of the physics of metal nanoparticles (NPs) is reported, followed by a more mathematical discussion of some specific subjects. Section 2.2 contains an overview on the most used methods to coat and functionalize AuNPs and, at the end of the chapter, in Section 2.3, the biocompatibility and organism biodistribution of AuNPs is presented. In this new edition, a more extended discussion is dedicated to the interactions between metal NPs and organisms.

2.1 Optical features

2.1.1 General description

Noble metal colloids are characterized by intense colors, resulting in light absorption and scatter in the visible region of the spectrum [3]. An early example of an application of this property is the rose window of the Notre Dame Cathedral in Paris, where silver and gold NPs are responsible for the colors of the glass. These effects are caused by one of the most important interactions of noble metal NPs with the electromagnetic (EM) field, the LSPRs (Figure 2.1).

Metals are characterized by the presence of "free" electrons and, when the diameter of metal nanostructures is in the 10–100 nm range, they interact with the light through [4]: (i) collective excitations of free electrons due to *intraband* transitions, giving rise to LSPR; (ii) transitions of electrons from occupied to empty bulk bands of a different index, called *interband* transitions; and (iii) surface dispersion or scattering of the free or unbound electrons, when their mean free path is comparable to the dimensions of the nanostructures. A *resonance* occurs when the frequency of an incident EM field matches the frequency of an intrinsic electronic oscillation. The collective and coherent oscillation of the electronic cloud of metals, called *plasmon*, causes a

https://doi.org/10.1515/9781501511455-002

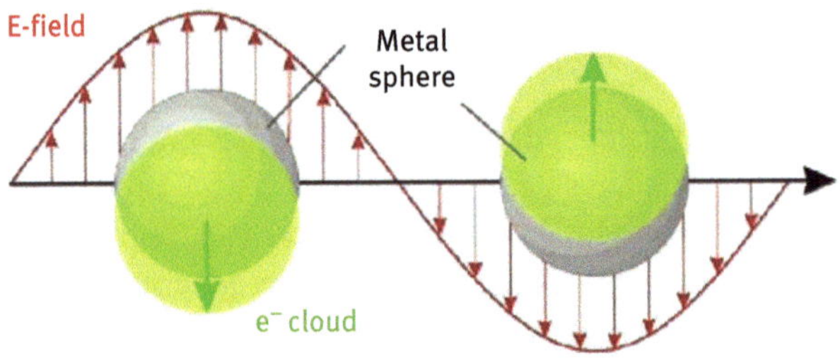

Figure 2.1: Scheme of a surface plasmon oscillation for a sphere, showing the displacement of the conduction electron charge cloud relative to the nuclei. Reproduced with permission from K. Kelly, E. Coronado, L. Zhao, and G. Schatz, Journal of Physical Chemistry B, 2003, 107, 668. ©2003, American Chemical Society [2].

displacement of the electrons from the nuclei, leading to the formation of various possible distributions in the nanostructure surface charges (i.e., dipole, quadrupole, and so on, see Figure 2.1). Each type of surface charge distribution is characterized by a specific resonance energy, the LSPR (Figure 2.1). When an incoming radiation of an appropriate frequency interacts with the nanostructure, its energy can be stored in the oscillation mode of the NP and can result in absorption and/or in light scattering. Noble metals such as copper, silver, and gold have strong visible-light plasmon resonances (Figure 2.2), whereas most other transition metals show only a broad and poorly resolved extinction band in the ultraviolet (UV) region [5].

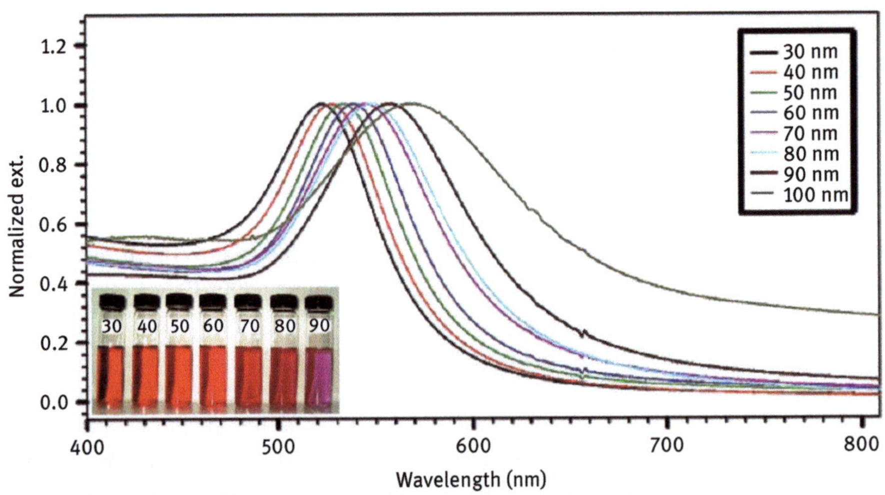

Figure 2.2: Normalized UV–Vis spectra for gold nanospheres (AuNSs) with different diameters in aqueous solution. Inset, photo showing the colors of gold solutions of nanospheres with different diameters from 30 to 90 nm. Reproduced with permission from P.N. Njoki, I-I.S. Lim, D. Mott, H-Y. Park, B. Khan, S. Mishra, R. Sujakumar, J. Luo, and C-J. Zhong, Journal of Physical Chemistry C, 2007, 14664. ©2007, American Chemical Society [6].

For noble metals, the presence of the LSPR in the visible region is attributed to the strong coupling between the plasmon resonance and *interband* excitations, since it can be assumed that: (i) the electrons of the conduction band of all metals can move freely and independently from the ionic background, and (ii) ions act only as scattering centers [7]. Therefore, the electron cloud of noble metals is higher polarizability than other transition metals, resulting in a shift of the plasmon resonance to lower (visible) frequencies with a characteristic sharp bandwidth (Figure 2.2) [8]. If each particle behaves independently with respect to the incident radiation, that is, in a dilute NP sample, the spectrum is composed from the sum of absorption and scattering modes. Therefore, the intensity of light transmitted can be described by eq. (2.1):

$$I = I_0 e^{\left[-\left(\sigma_{\text{abs}} + \sigma_{\text{sc}}\right)NL\right]} \tag{2.1}$$

where I_0 is the intensity of the incoming light, N is the number of particles per unit volume, and L is the length of the optical path in the sample. The quantity $\sigma_{\text{ex}} = \sigma_{\text{abs}} + \sigma_{\text{sc}}$ is also known as the extinction cross section, while σ_{abs} and σ_{sc} are, respectively, the absorption and scattering cross section of the NP. It is worth to notice that the LSPR position and width are influenced by the size and shape of the particles, the dielectric function of the medium, and the presence of other nanostructures in close proximity (Figures 2.2 and 2.3) [1].

For metal nanospheres (NSs), *interband* electronic transitions are not very sensitive to particle size (except in the case of sub-2 nm metal clusters, which are made of a few atoms), and are located at high energy (UV region of the spectra). For NPs with diameters between 10 and 30 nm, the dominant effect in the visible region is the excitation of plasmon modes. In this size range, and in the simple case of spherical NPs, a single dominant plasmon mode of a dipolar nature is excited; for gold (Figure 2.2) this mode is at about 515–520 nm, and for silver at 400 nm. However, scattering effects are more important for NSs with a diameter of more than 30 nm, where electrons are accelerated by the EM field and radiate energy in all directions. Thus, electrons lose energy by a damping effect on their motion. Indeed, NPs spectrum is less intense, wider, and red-shifted when the particle size increases (Figure 2.2) [4]. Overall, the depolarization field term causes the shift to larger wavelengths while the radiation damping causes the intensity decreasing and widening of the spectrum [10]. Finally, scattering effects dominate the response of NSs with diameters larger than 100 nm and higher order modes, that is, quadrupolar and octupolar, contribute to the interaction between light and matter. In a theoretical/experimental work on spherical AuNPs by El-Sayed and coworkers, the sum of the effects discussed above results in a red shift on the λ_{max} of the LSPR of about 0.7 nm for every 1 nm increase in particle radius (for diameter > 25 nm) [11]. Interestingly, λ_{max} is almost independent from the particle size when particle diameter is smaller than 25 nm.

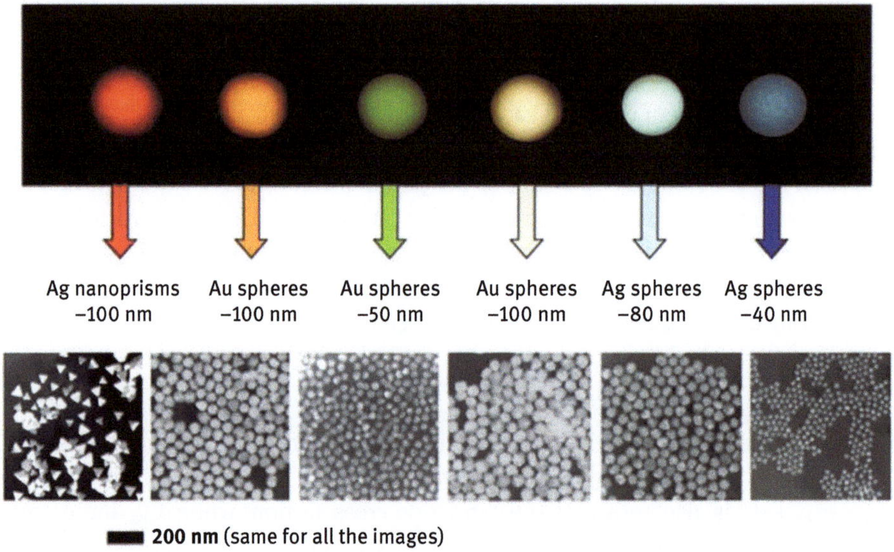

Figure 2.3: Size, shape, and composition of metal nanoparticles can be systematically varied to produce materials with distinct optical properties. The upper panel shows the dark field analysis of the nanoparticles shown in the bottom panel. Reproduced with permission from N.L. Rosi and C.A. Mirkin, Chemical Reviews, 2005, 105, 1547. ©2005, American Chemical Society [9].

The correlation between NS size and λ_{max} of the LSPR band was theoretically described for the first time by Mie [12]. He solved the Maxwell's equations in the quasistatic regime, that is, he assumed that the field was constant throughout the particle albeit time/frequency dependent, and in the dipole approximation, that is, NSs smaller than the incident wavelength, obtaining eq. (2.2):

$$\frac{\sigma_{ex}}{1,000 \cdot V} = \frac{18\pi\varepsilon_a^{3/2}}{\lambda} \cdot \frac{\omega\varepsilon_2(\omega,\ R)}{(\varepsilon_1(\omega,\ R) + 2\varepsilon_a)^2 + \varepsilon_2^2(\omega,\ R)} \tag{2.2}$$

where σ_{ex} is the extinction cross section, ω is the angular frequency, V is the volume of each sphere, ε_a is the medium dielectric constant, and ε_1 and ε_2 are the real and complex parts of the dielectric function of the metal [12]. The resonance condition is roughly fulfilled when $\varepsilon_1(\omega, R) = -2\varepsilon_a$ if ε_2 is small or weakly dependent on ω. Equation (2.2) explains the dependence of LSPR on the dielectric function of the surrounding medium ε_a [6]. In this model, the dependence of the LSPR band position to the size of NSs is associated with the dependence of the refractive index of NPs to R [13]. Therefore, an intrinsic dependence of the real and imaginary parts of the dielectric function of metals to R is indicated in eq. (2.2) [1]. Interestingly, the size dependence is lost if the dielectric constant of the bulk metal is used to solve Maxwell's equations. It is important to note (Figure 2.4) that: (i) there is a direct dependence between the NSs extinction cross section and the sphere volume, and (ii) the σ_{ex} of gold NSs is typically four to five orders

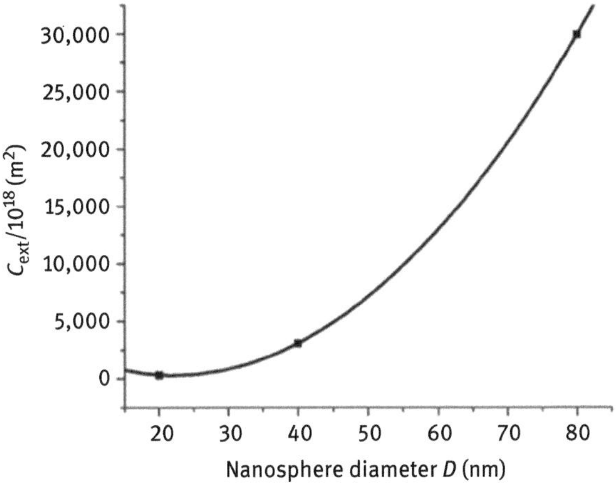

Figure 2.4: Variation of extinction cross section (C_{ext}) with nanosphere diameter. Reproduced with permission from P.K. Jain, K.S. Lee, I.H. El-Sayed and M.A. El-Sayed, Journal of Physical Chemistry B, 2006, 110, 7238. ©2006, American Chemical Society [14].

of magnitude higher compared to those of organic dyes [14]. At the same time, the relative contribution of scattering to the total extinction cross section (C_{ext}) increases with the square of particle volume as shown in Figure 2.5.

The trend of the scattering to absorption ratio associated with the NP volume is caused by an increase in radiative damping for larger particles [14]. Overall, the LSPR features of AuNSs with diameter >20 nm are mainly exploited for imaging applications, among which selective scattering imaging in dark field (DF) microscopy and confocal microscopy [15, 16], while AuNSs with diameter in the 3–10 nm range are generally suggested as photoabsorbers for laser photothermal therapy and absorption contrast imaging (Chapter 3) [17–20].

Beyond size, the shape of metal NPs has a striking influence on optical properties (Figure 2.3). For example, the surface plasmon absorption maximum (λ_{max}) of elongated AuNPs strongly depends on their aspect ratio r, that is, the length-to-width ratio, as explained by Gans (eq. (2.3)) [21, 22]:

$$\lambda_{max} = 420 + 95r \quad (2.3)$$

In general, the main LSPR redshifts with r (Figure 2.6).

In rod- or ellipsoidal-shaped NPs, the LSPR splits to two modes corresponding to the oscillation along and perpendicular to the long axis of the particle (Figure 2.6) [21, 23]. Since the shape affects the electron charge density on the particle surface, other shapes of AuNPs such as triangle, cube, or shell exhibit a redshifted LSPR band compared to their spherical analogs [21, 24]. Redshift of LSPR to the near-infrared window (650–900 nm) is especially desirable for in vivo applications because the tissue absorption is minimal in that range, improving the light penetration [20, 25].

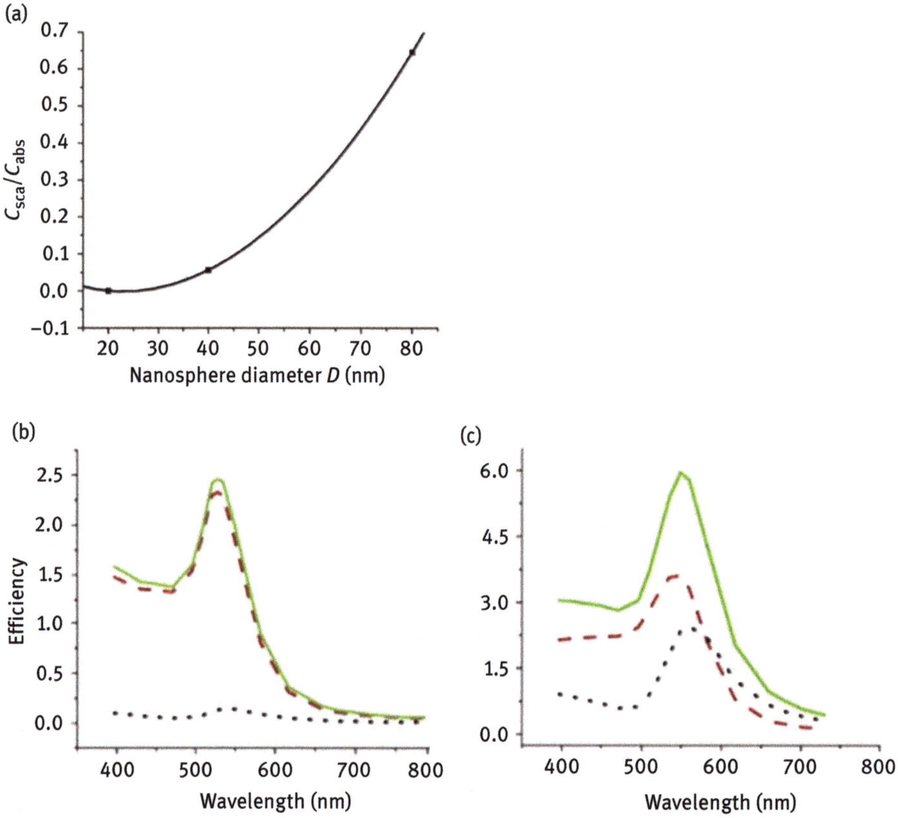

Figure 2.5: (a) Variation of the ratio between scattering and absorption cross sections (C_{sca}/C_{abs}) with nanosphere diameter D. (b, c) Calculated spectra of the efficiency of absorption Q_{abs} (- - -), scattering Q_{sca} (. . . .), and extinction Q_{ext} (-) for gold nanospheres of diameter (b) $D = 40$ nm, (c) $D = 80$ nm. Reproduced with permission from P.K. Jain, K.S. Lee, I.H. El-Sayed and M.A. El-Sayed, Journal of Physical Chemistry B, 2006, 110, 7238. ©2006, American Chemical Society [14].

The LSPR can be also redshifted by maintaining in close proximity a number of NPs [26, 27]. Here, this approach was recently employed (Chapter 3) to bring again the potentiality of AuNPs to the forefront of cancer treatments [28–30].

The dipoles of two or more metal NPs couple when in close proximity, resulting in a shift of the LSPR mode (Figures 2.7 and 2.8).

For example, aqueous solutions of single 10 nm AuNSs have a typical plasmon extinction maximum at 520 nm, while if aggregated in a both controlled or noncontrolled manner, a redshift and widening in the extinction band is observed [26, 27, 31].

This effect was investigated both theoretically and experimentally for fixed (Figure 2.8) and nonfixed distances (Figure 2.7) [15, 33, 34]. The magnitude of the assembly-induced plasmon shift depends on the strength of the interparticle coupling,

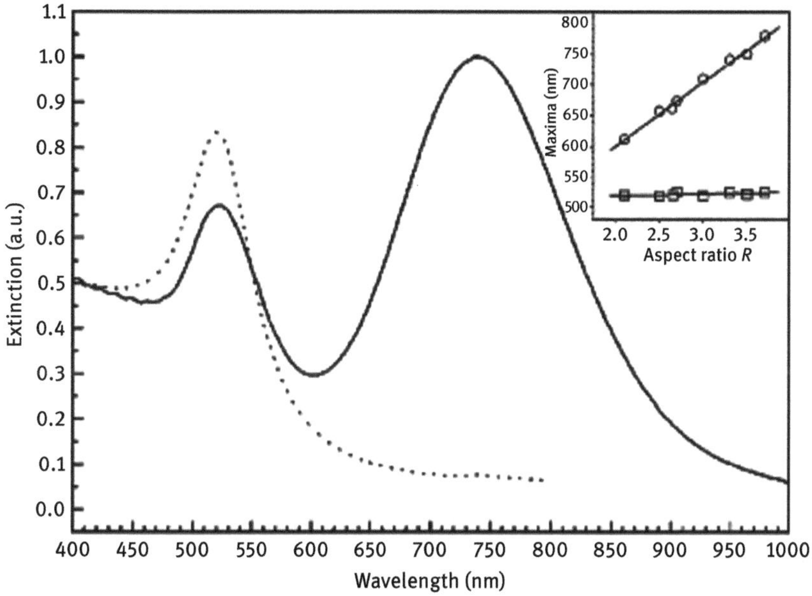

Figure 2.6: Extinction spectrum of a sample consisting of a colloid of nanorods having an aspect ratio $r = 3.3$ and a transversal dimension of 22 nm (solid line), compared to one of 22 nm nanospheres (dotted line). The inset shows how the maxima of the transverse (squares) and longitudinal (circles) surface plasmon modes vary with the aspect ratio. Reproduced with permission from X. Huang, S. Neretina and M.A. El-Sayed, Advanced Materials, 2009, 21, 4880. ©2009, Wiley-VCH [21].

which, in turn, depends on the distance between the individual NPs. Therefore, the plasmon shift can provide a measure of the distance between pairs of NP [15]. El-Sayed and co-workers proposed an empirical equation (eq. (2.4)) in order to estimate the interparticle distance from experimentally observed plasmon shifts [31, 35]:

$$\frac{\Delta\lambda}{\lambda_0} \approx A \cdot e^{\left(\frac{-(s/D))}{B}\right)} \tag{2.4}$$

where $\Delta\lambda/\lambda_0$ is the fractional plasmon shift, s is the interparticle edge-to-edge separation, D is the particle diameter, and A and B are two adimensional parameters typical of the experimental setup. This equation was deduced for pairs of coupled gold NPs (in 20–100 nm diameter range), in protein medium, at fixed distances, and by DF imaging experiments illuminating with unpolarized white light. In this setup, A and B parameters are estimated to be, respectively, 0.18 and 0.23 [35]. Interestingly, the optical analysis of dimers of AuNPs may result as an alternative to the Förster resonance energy transfer (FRET) for single-molecule experiments, especially for applications demanding long observation times (seconds to hours) or large distances (usually up to 2.5 times the diameter of the spheres). In FRET, the

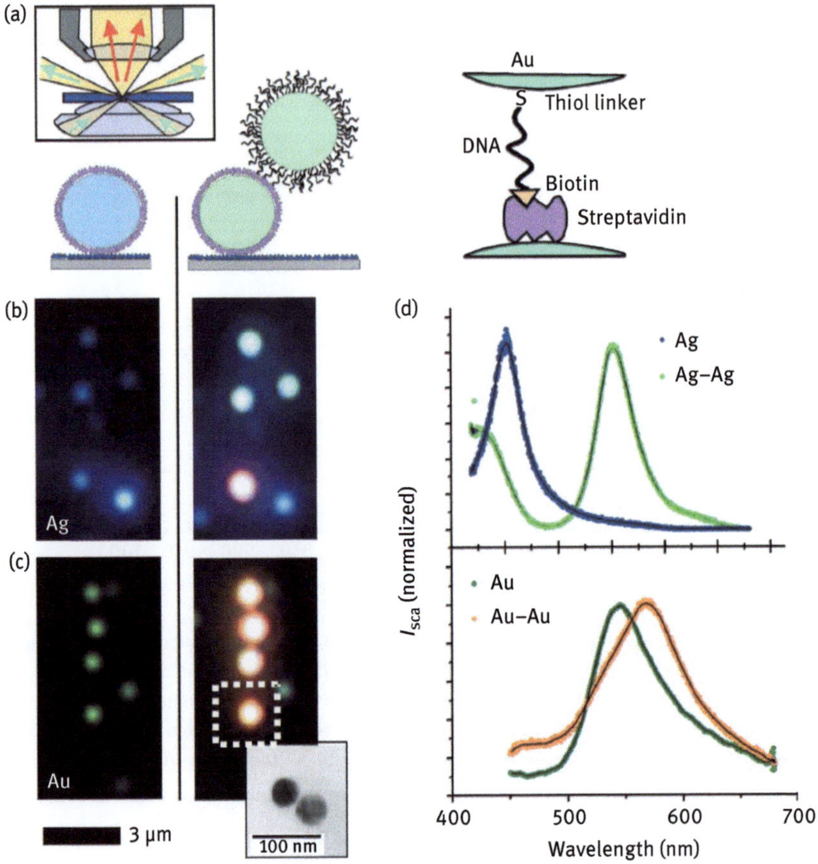

Figure 2.7: Effect of the coupling of DNA-functionalized gold and silver nanoparticles on their color when observed by dark field microscopy. (a) Two gold or silver nanoparticles can be linked together through a biotin–streptavidin interaction. Inset: principle of transmission dark field microscopy. (b) Single silver particles appear blue (left) and particle pair blue-green (right). The orange dot in the bottom is an aggregate of more than two particles. (c) Single gold particles are green (left) while pairs are orange (right). Inset: representative transmission electron microscopy (TEM) image of a particle. (d) Representative normalized scattering spectra of single particles and particle pairs for silver (top) and gold (bottom). Silver particles show a larger spectral shift (102 nm) than gold particles (23 nm), together with a stronger light scattering and a smaller plasmon line width. However, gold is chemically more stable and is more easily conjugated to biomolecules via –SH, $–NH_2$, or –CN functional groups. Reproduced with permission from C. Sonnichsen, B.M. Reinhard, J. Liphardt and A.P. Alivisatos, Nature Biotechnology, 2005, 23, 741. ©2005, Nature [15].

observation of the fluorescence of a single-organic dye is often hindered by blinking effect and/or rapid photobleaching phenomena, limiting the continuous observation time to a few tens of seconds. Furthermore, discriminating variations in relative dye orientation with respect to changes in distance may sometimes result difficult [36].

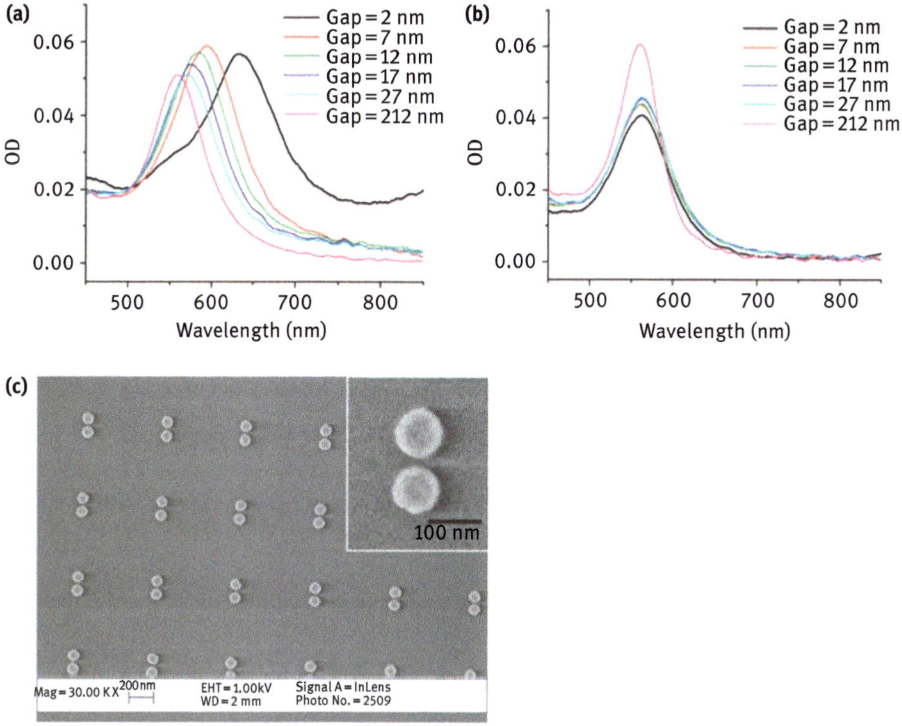

Figure 2.8: Microextinction spectra of Au nanodisk pairs at various interparticle gaps for incident light polarization direction (a) parallel and (b) perpendicular to the interparticle axis. OD (optical density) = $-\log_{10}(T)$, where I is the local light transmittivity. (c) SEM image of an array of nanodisk pairs used to determine the *plasmon ruler equation*; in this image each nanodisk has a diameter of 88 nm, a thickness of 25 nm, and an interparticle edge-to-edge separation gap of 12 nm. Reproduced with permission from P.K. Jain, W.Y. Huang and M.A. El-Sayed, Nano Letters, 2007, 7, 2080. ©2007, American Chemical Society [32].

The plasmon resonance signals, instead, neither blink nor bleach and do not depend to the relative probe orientation when exploited with nonpolarized light [35]. It is worth to remember that plasmon shifts in experiments exploited with polarized white light depend on the orientation of the EM field (Figure 2.8). Furthermore, gold and silver NPs are more stable under physiological conditions and under laser illumination than organic dyes. The range of distances accessible with plasmon coupling in a pair of NPs depends on the size and coating of the particles. In general, the accessible distance range (≈10–200 nm) is larger than with FRET (2–8 nm) [36], and particle separations of up to 70 nm have demonstrated about 1 nm resolution (conditions: in vitro experiment, 40 nm NPs, and 0.1 nm spectral resolution for determining the plasmon resonance position) [33]. On the other hand, the AuNPs diameter required for this application should be over 20–30 nm in order to ensure the collection of scattering signals [14, 15], which may affect the structural conformation and the activity of many targets. Furthermore, exploitation of NPs for single molecule imaging in living cell experiments may result very difficult because of the high scattering background and the required selective interactions.

In addition to the phenomena mentioned previously, the excitation of the LSPR may result in a photoluminescence (PL) emission from sharply angled nanomaterials (*lightning rod effect*). The quantum efficiency (the number of photons emitted over the number of absorbed photons) of the PL is very low for bulk noble metals, typically of the order of 10^{-10} [37]. On the other hand, due to the *lightning rod effect*, the PL efficiency of gold nanorods increases by six orders of magnitude from bulk, while for gold nanocubes reaches 10^{-2}, about 200 times higher than that of gold nanorods [38, 39]. PL was not recognized for 15 nm spherical NPs, while it was found, and it is easily tunable, for very small gold clusters (<5 nm) [40–42]. The origin of the PL was attributed to recombination of electron–hole pairs. The electrons from the filled *d* band of the electronic states are promoted to the *sp* conduction band above the Fermi level by incident photons. Both electrons and holes relax by scattering with phonons resulting in the emission of light. Noticeably, the PL signals are intense enough, due to the drastic σ_{ex} of the gold nanostructures, to be exploited in single molecule experiments and to engineer multicomponent nanosystems for imaging or theranostic applications (Chapter 3).

In the following section, a more mathematical description of the physical properties of AuNPs is reported.

2.1.2 Introduction to the dielectric function

The dielectric function (ε) is, in general, a physical quantity that describes how an EM field and a dielectric medium affect each other, within the linear region. In particular, the dielectric function is a quantity that measures the possibility of a material to polarize itself in the presence of an electric field and depends on both the angular frequency ω and the wavenumber k. In a metal, the dielectric function ε results from the contribution of both the electrons near the Fermi level, which are essentially "free," and the bound electrons. The contribution of free electrons to ε arises from intraband transitions, named ε_{intra}. On the other hand, bound electrons in fully occupied bands can be promoted to nonfull bands after the interaction with photons at peculiar energy. Full and nonfull bands are separated by an energy gap; hence, bound electrons are promoted via interband transitions. The contribution of these electrons to ε can be named ε_{inter}. If ε_{exp} is the measured bulk dielectric constant, $\varepsilon_{exp} = \varepsilon_{inter} + \varepsilon_{intra}$. The dependence of ε_{intra} on the frequency ω is adequately described by the *Drude model* [7, 43]. In this model, metals are approximated to a plasma, that is, a medium composed of an equal number of conduction band electrons and of positively charged ion cores, in which the conduction electrons are not bound to atoms. The contribution of conduction electrons to the dielectric function is approximated as follows:

$$\varepsilon_{\text{intra}}(\omega) = 1 - \frac{\omega_{\text{P}}^2}{\omega(\omega + i/\tau)} \tag{2.5}$$

where τ is a relaxation time of plasma oscillation, and ω_{p} is the plasma frequency, defined by $\omega_{\text{p}} = 4\pi\, ne^2/m$, where n is the density of conduction electrons, while m and e are the mass and the charge of the electron, respectively. The physical meaning of the plasma frequency is described in the following. A Coulomb force is produced in a plasma when all conduction electrons are forced to one direction with respect to the positive ion lattice by the application of an external electric field. When the field is removed and, if no other forces are acting on the system, the whole plasma oscillates at a frequency equal to ω_{p} [7]. The damping factor $1/\tau$ in the Drude model is the inverse of the free electrons relaxation time τ and arises from various scattering processes of electrons, such as electron–electron and electron–phonon scattering.

In >20 nm metal NPs, an additional size-dependent damping term, which is related to electron scattering from the boundary of the particle, appears. In the nanosize regime, the contribution of free electrons to the dielectric function by considering the surface damping term is described in the following equation [36]:

$$\varepsilon_{\text{intra}}(\omega,\ R) = 1 - \frac{\omega_{\text{P}}^2}{\omega(\omega + i/\tau + i\omega_{\text{D}})} \tag{2.6}$$

Notice that $\omega_{\text{D}} = A \cdot v\,F/R$ is a frequency damping term that is directly proportional to the Fermi velocity v_{F}, that is, the Fermi velocity associated with the Fermi energy E_{F} via $E_{\text{F}} = 1/2mv_{\text{F}}{}^2$, and it is inversely proportional to the particle radius R. F is the Faraday constant (96.485 C/mol) and A is a constant of proportionality.

The values of $\varepsilon_{\text{inter}}$ can be estimated for several metals at all wavelengths because the dielectric function ε_{exp} can be measured while $\varepsilon_{\text{intra}}$ can be described using the *Drude model*. In Figure 2.9, the real part ($\varepsilon_{\text{real}}$) and the imaginary part (ε_{im}) of the dielectric function for copper, gold, and silver are reported [10].

Both the real and the imaginary components are decomposed in their interband and intraband contributions. The major contributions of the interband transitions to the imaginary component are below 325 nm for silver, 550 nm for gold, and 600 nm for copper. On the other hand, the contribution of the interband transitions to the real part is only a positive and almost constant background. Indeed, interband transitions contribute to absorption but not to plasmon modes.

2.1.3 Plasmonic properties of small spherical metal nanoparticles

Metal NPs have attracted great attention for their peculiar interaction with light driven by the LSPRs. In general, a resonance occurs when a harmonic system with an intrinsic frequency interacts with an external periodic force with a frequency very close to it.

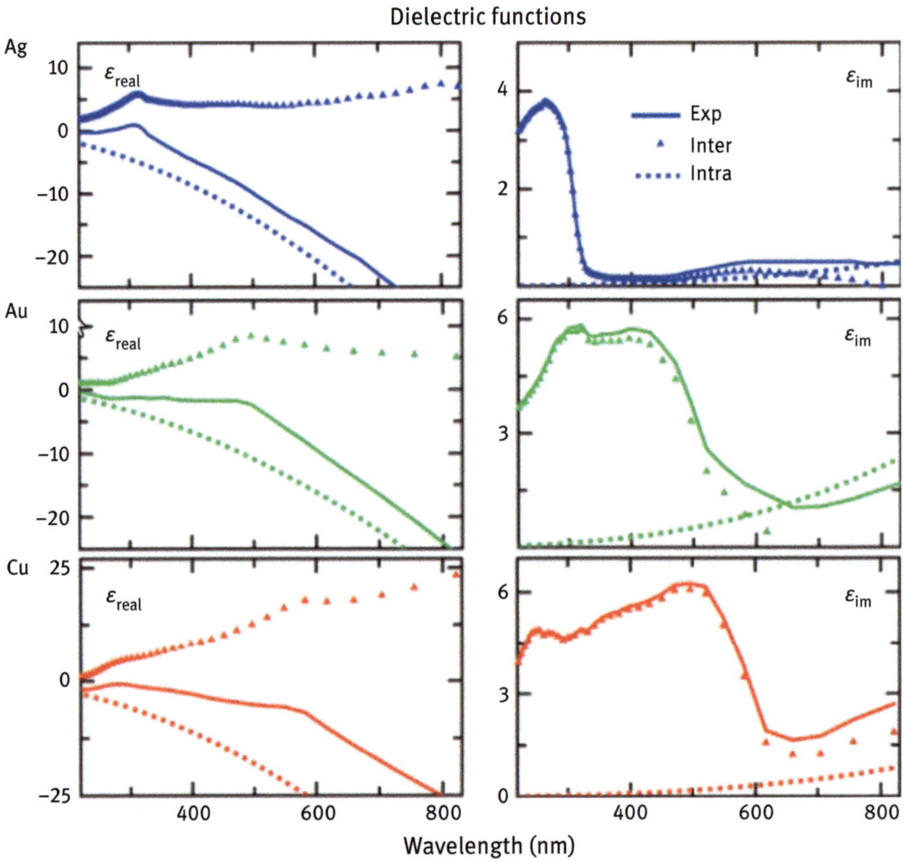

Figure 2.9: Solid lines: real (left) and imaginary (right) parts of the dielectric function for silver (blue), gold (green), and copper (red). In each panel, the interband (triangles) and intraband (dots) contributions are shown. Reproduced with permission from R. Krahne, G. Morello, A. Figuerola, C. George, S. Deka and L. Manna, Physics Reports, 2011, 501, 75. ©2011, Elsevier [4].

This interaction causes a big transfer of energy between the exciting system and the resonator. For metal NPs, the resonator is the charge density oscillation, that is, the plasmon, in the nanostructure and the exciting system is the light.

In principle, the interaction of metal NPs with light can be described by solving the Maxwell's equations with specific boundary conditions. Mie solved the Maxwell's equations for spherical NPs with a dielectric function ε (the dependence on ω and k is implicit) in a uniform medium of dielectric function ε_{medium}, and subjected to a monochromatic radiation field with a plane waveform [12]. The analytical solutions are cross sections σ, that is, series whose terms contain the various multipolar contributions for both extinction (eq. (2.7)) and scattering (eq. (2.8)) features:

$$\sigma_{ext} = \frac{2\pi}{|k|^2}\sum_{l=1}^{\infty}(2l+1)\mathrm{Re}\{A_l + B_l\} \tag{2.7}$$

$$\sigma_{SC} = \frac{2\pi}{|k|^2}\sum_{l=1}^{\infty}(2l+1)\left\{|A_l|^2 + |B_l|^2\right\} \tag{2.8}$$

In the above equations, $k = 2\pi(\varepsilon_{medium})^{1/2}/\lambda$ is the magnitude of the wave vector of the incoming radiation, and A_l and B_l are the electric and magnetic scattering coefficients for the multipolar mode defined by l ($l = 1$ for dipole, 2 for quadrupole, 3 for octupole, etc.). The coefficients are explicitly expressed as Riccati–Bessel functions, and depend on ε, ε_{medium}, and on the product of the particle radius R with k. Equations (2.7) and (2.8) are a satisfactory description of the optical behavior of metal NPs in a broad range of sizes. The ε is often obtained from experiments or model calculations since it may differ from its bulk value at nanosize regime and may depend on R. Overall, eqs. (2.7) and (2.8) may result in difficult interpretation and visualization; hence, approximate solutions are often considered in various limiting cases.

One of these approximations is the quasistatic regime (also called "electrostatic" or "nonretarded"), in which the field applied to the particle is assumed to be constant throughout the solid, albeit still time/frequency dependent. The quasistatic regime is valid in the dipole approximation, that is, for NSs much smaller than the wavelength of the incident light and with radii smaller than ~30 nm. Furthermore, the speed of light is considered infinite, in order to avoid the coupling between electric and magnetic fields. Thus, the optical response of the NP is basically dictated by the electric field, and the NP polarizability can be described as follows:

$$\alpha = 4\pi R^3 \varepsilon_{medium} \frac{\varepsilon - \varepsilon_{medium}}{\varepsilon + 2\varepsilon_{medium}} \tag{2.9}$$

The resonance is reached when $\varepsilon \sim -2\varepsilon_{medium}$. In this case, the incident light uniformly polarizes the NP and the charges are homogeneously displaced, that is, free electrons move in phase. Therefore, a dipolar charge distribution is generated on the particle surface, while higher order electric multipoles are negligible.

For particles smaller than 10 nm, the scattering cross section σ_{sc} is much smaller than the absorption cross section σ_{abs}. Indeed, scattering is a radiative process and requires the coupling between the electric and magnetic fields, that is, negligible in this size regime. Therefore, the extinction cross section of a single particle is almost the same of the absorption cross section because $\sigma_{ext} = \sigma_{sc} + \sigma_{abs}$ and can be approximated as follows [44]:

$$\sigma_{ext} \approx \sigma_{abs} = \frac{24\pi^2 R^3 \varepsilon_{medium}^{3/2}}{\lambda} \frac{\varepsilon_{im}}{(2\varepsilon_{medium} + \varepsilon_{real})^2 + \varepsilon_{im}^2} \tag{2.10}$$

where R is the radius of the particle, and ε_{real} and ε_{im} are, respectively, the real and imaginary parts of the dielectric function, which depend on the frequency ω of the metal. It is worth to remember that ε_{im} is generally small compared to ε_{real} and does not vary much with ω (or with the wavelength), as shown in Figure 2.9. Roughly, the maximum in absorption is reached for $\varepsilon_{real} \equiv \mathrm{Re}(\varepsilon(\omega)) = -2\varepsilon_{medium}$ (also known as the Fröhlich condition), while the peak width is dependent on ε_{im} and ε_{medium}. Interestingly, the size of the particle seems to have a small influence on the peak position and on the width of the plasmon band. Indeed, in metal NPs smaller than 20 nm in diameter, an additional size-dependent damping term in the dielectric function has to be considered. This term is related to electron scattering from its boundary. The free electron contribution to the dielectric function considering the surface damping term is in eq. (2.11) [45]:

$$\varepsilon_{intra}(\omega,\ R) = 1 - \frac{\omega_p^2}{\omega(\omega + i/\tau + i\omega_D(R))} \tag{2.11}$$

where $\omega_D(R) \propto V_F/R$ is a frequency damping term that is directly proportional to the Fermi velocity v_F (that is associated with the Fermi energy E_F via $E_F = 1/2\ m{v_F}^2$) and inversely proportional to the particle radius R. This effect, called "nonlocal (or surface) dispersion" or "surface damping," does not influence the peak position of the resonance band. Instead, it leads to a broadening of the plasmon band, because the electrons lose coherence in their motion at each scattering event with the surface of the NP. The broadening of the plasmon band in small particles is a very relevant effect; for example, the LSPR band practically disappears for particle sizes smaller than 2 nm. Here, it is interesting to mention that Baida and coworkers recorded the optical extinction coefficient of individual silver NPs in the size range between 10 and 50 nm, observing that the plasmon band starts broadening below 25 nm due to surface damping [46].

2.1.4 Plasmonic properties of large spherical metal nanoparticles

In spherical NPs, higher order multipolar modes become nonnegligible starting from particle radii around 30 nm, where quadrupole modes start appearing. In this section, the first higher order multipolar modes are considered. In agreement to Krahne, the general resonance conditions for multipolar modes in the "quasistatic" regime occur for $\varepsilon = -\varepsilon_{medium}(l + 1)/l$. Thus, the resonance condition for a quadrupolar mode is $\varepsilon = -(2/3)\varepsilon_{medium}$, and it is usually shifted at lower energies compared to the dipolar mode [4]. The quadrupolar mode is initially visible in the optical extinction spectrum of NPs as a small shoulder at a lower wavelength of the dipolar LSPR mode. As the NP size increases, the contribution from the quadrupolar mode increases. The visibility of the quadrupolar mode also depends from the metal. For example, the quadrupolar mode of small AuNPs is weak and hidden by the interband transitions [10]. On the other hand, in silver NPs such overlap does not occur and, therefore, the quadrupolar mode is already evident at smaller sizes than found with gold [10]. Octupolar modes are relevant only for much bigger particles, and for even higher order modes. The resonance conditions are fulfilled when $\varepsilon = -\varepsilon_{medium}$, which is the resonance condition of the planar surface plasmon, that is, the plasmon resonance for a flat metal surface in contact with a dielectric medium of dielectric function ε_{medium}. It is worth to remember that in multipolar modes the short charge waves prevent the distant charges interactions, and, thus, each small region on the surface of the metal behaves as in a planar bulk metal.

Moreover, the optical behaviors of high-order multipolar modes are also influenced by retardation effects, as these modes are active in NPs larger than 30 nm in diameter. An example is the "energy-shifting" effect, which arises because the particle size is no longer negligible compared to the wavelength of the radiation. In particular, this effect appears when the diameter $2R$ of the particle is around 1/10 of the mode wavelength $\lambda_m = \lambda/(\varepsilon_{medium})^{1/2}$ of the radiation in the medium surrounding the particle. In this case, the wavelength of the incoming radiation cannot be considered as infinite and the field is not homogeneous everywhere inside the particle. Indeed, when a dipole mode is excited in a spherical particle, the distance between opposite charges is the NP diameter. Thus, changes occurred at a side of the particle have a phase retardation equal to $4\pi R/\lambda_m$ on the opposite side. The oscillation period of the dipole mode increases in order to take such retardation into account. Hence, the energy of the plasmon peak associated with a dipole mode redshifts as the particle size increases. When higher order multipolar modes are involved, the distance between opposite charges on the surface of the particle is smaller than the particle diameter, and the phase retardation is smaller than in the dipolar mode. Indeed, the phase retardation in a spherical metal particle scales roughly as $4\pi R/(l \cdot \lambda_m)$ with $l = 1, 2, 3$, for dipole, quadrupole, and octupole modes. It is interesting to notice that dipole modes are more influenced by the particle size than the higher order modes due to retardation effects. Another retardation effect that is important for 20 nm NPs and dominant for 100 nm NPs is related to radiation scattering. During irradiation, electrons are accelerated as a consequence of the applied EM field, and particles lose energy by emitting radiations. Briefly, part of the energy of the plasmonic oscillation is converted into photons. This phenomenon leads to a broadening of the LSPR band in the extinction spectrum, resulting in an apparent decrease in its intensity with respect to the background (that is mostly given, at the lowest wavelengths, by nonresonant scattering). Overall, in large particles the interaction with incident light is dominated by radiative processes, that is, the photons are scattered, as opposed to small particles. Meier and Wokaun calculated the consequences of the retardation effects on the extinction efficiency of NP by applying the "modified long wavelength approximation" approach that has described both the "energy-shifting" effect of the depolarization field inside NPs and the broadening of the band due to radiation damping [47].

2.1.5 Surface-enhanced/quenched fluorescence

When dyes or quantum dots are near to the surface of metal NPs, their fluorescence can be quenched or enhanced. The fluorescence is usually quenched when the fluorophore is directly adsorbed on the particle surface, and, vice versa, enhanced if positioned few nanometers away. Fluorescence enhancing/quenching is attributed to several reasons, such as (i) increased absorption by the fluorophore, (ii) faster radiative decay rate of the fluorophore, and (iii) better coupling of the emission to the far field. Briefly, metal NPs act as nanoantennas that collect, concentrate, and transfer to dyes the energy of the incident light due to the "field enhancement effect" [48]. The interactions are maximized when the plasmon resonance of the NP overlaps the absorption band of the dye. The frequency of the LSPR mode of metal NPs can be tuned over a wide spectral range. This behavior has resulted in several interesting applications for (bio)sensing (Chapter 3).

When an isolated fluorophore is weakly excited by an electric field E_0 (measured at the fluorophore position), the fluorescence S_0 is described as follows:

$$S_0 = \xi \eta_0 |d \cdot E_0|^2 \tag{2.12}$$

where d is the dipole moment, ξ is the collection efficiency, and η_0 is the quantum efficiency of the fluorophore. The quantum efficiency is defined as follows:

$$\eta_0 = \frac{\Gamma^0_{\text{rad}}}{\Gamma^0_{\text{rad}} + \Gamma^0_{\text{nrad}}} \tag{2.13}$$

$\Gamma^0{}_{\text{rad}}$ and $\Gamma^0{}_{\text{nrad}}$ are the radiative and nonradiative decay rates, respectively. When the fluorophore is in the proximity of the metal NP: (i) the local field E is stronger, (ii) a new radiative decay rate Γ_{rad} arises due to the coupling of the electrons of the fluorophore and of the NP, and (iii) an additional nonradiative decay channel with decay rate Γ_{nrad} occurs due to absorption by the metal. The fluorescence signal enhancement is defined as follows:

$$\frac{S}{S_0} = \frac{\eta}{\eta_0} \cdot \frac{|d \cdot E|^2}{|d \cdot E_0|^2} \tag{2.14}$$

And then the modified quantum efficiency η is described by eq. (2.15):

$$\eta = \frac{\eta_0}{\frac{(1-\eta_0)}{F} + \frac{\eta_0}{\eta_a}} \tag{2.15}$$

$F = \Gamma_{\text{rad}}/\Gamma^0{}_{\text{rad}}$ is the Purcell factor and describes the enhancement in radiative decay, while $\eta_a = \Gamma_{\text{rad}}/(\Gamma_{\text{rad}} + \Gamma_{\text{nrad}})$ is the "antenna efficiency" [49]. If the quantum efficiency η_0 of the fluorophore is 1, then $\eta = \eta_a \leq 1$ in any case. Thus, the quantum efficiency can be only reduced when the fluorophore is close to the particle. However, when

$\eta_0 < 1$, as often happens, η depends on both F and η_a. When both these two parameters are large, η is maximized. Therefore, in order to strongly improve the fluorescence, as also described in eqs. (2.14) and (2.15), the system has to (i) locally generate a strong electric field and (ii) demonstrate a large antenna efficiency and the Purcell factor.

2.1.6 Surface-enhanced Raman scattering

The surface-enhanced Raman scattering (SERS) effect is related to a molecule in close proximity to a metal NP or to a roughened metal surface. The effect was discovered in 1977, described in 1978, and since then has been exploited in various applications, including (bio)sensing [50–52]. As for enhanced fluorescence, the enhancement of the Raman scattering is mainly due to the increase of the EM field near the surface of metal NPs. For a spherical NP in a medium with a dielectric constant $\varepsilon_{\text{medium}}$, the absolute square of the field E outside the sphere is expressed by eqs. (2.16) and (2.17) by assuming that the light incident on the particle has an EM field vector E_0 aligned along one of the axes of the particle (the z-axis):

$$|E|^2 = |E_0|^2\left[|1-g|^2 + 3\cos^2\theta\left(2\mathrm{Re}(g) + |g|^2\right)\right] \tag{2.16}$$

$$g = \frac{\varepsilon - \varepsilon_{\text{medium}}}{\varepsilon + 2\varepsilon_{\text{medium}}} \tag{2.17}$$

In eq. (2.16), θ is the angle between the excitation field vector and the molecule on the surface of the particle. The Raman intensity depends on $|E|^2$. The term g becomes large when resonance conditions are met, and in such a case, eq. (2.16) can be approximated as eq. (2.18):

$$|E|^2 - |E_0|^2|g|^2\left[1 + 3\cos^2\theta\right] \tag{2.18}$$

The highest Raman intensity happens when the molecule on the surface of the particle is along the z-axis of the particle, either at 0° or at 180°. Due to the field enhancement, the probability that a molecule interacts with the incident light and scatter a Raman photon is increased. In addition, scattered Raman photons may also excite the surface plasmons of the particle, which radiate at the same frequency. These photons may locally produce an enhanced oscillating electric field at the Raman frequency, which, in turn, increase the probability of stimulated Raman scattering. Therefore, the intensity of the Raman signal can be enhanced by two processes, considering the incident and the scattered fields, and the overall enhancement can be written as follows:

$$E_r \propto \frac{|E|^2|E'|^2}{|E_0|^4} \approx 16|g|^2|g'|^2 \tag{2.19}$$

where the variables with apex refer to the re-radiated fields. In the case of a prolate spheroid and assuming that the incident electric field is aligned with the long axis of the spheroid, the enhancement is maximal at the tip regions of the spheroids, that is, where the curvature is highest. Moreover, more prolate is the spheroid and bigger is the enhancement [53]. However, the maximum Raman enhancement occurs only if all the molecules are located at the apex regions. In real cases, molecules are located on all the surface of the particles. Thus, the field enhancement is an average described by $4|g|^2|g'|^2$ for a sphere, which decreases to zero for infinitely prolate ellipsoids while it is described by $\sim 6|g|^2|g'|^2$ for infinitely oblate ellipsoids [53]. It is worth to notice that in the above descriptions the dielectric constant of the metal is considered as in the Drude model, without considering other terms. The magnitude of SERS depends on the shape of the particle as well as on its overall size. Factors that tend to reduce the field enhancement, that is, SERS, are: (i) the scattering of electrons from the surface for small particles (Section 2.1.3) and (ii) the radiation damping for big particles (Section 2.1.4). Overall, by considering most of the light–matter interactions, a huge range of NPs have demonstrated SERS maximization [53]. In the above description, the discussion has been focused on the interaction between light and a single NP. On the other hand, the SERS enhancement is dramatically increased when two or more NPs interact, producing "hot spots." Recently the development of effective SERS tools has attracted increasing attention as SERS is an interesting and promising topic. Interested readers are encouraged to refer elsewhere for some focused reviews [54–56].

2.2 Coatings for AuNPs

It is worth to notice that while the optical and physical properties of AuNPs can be directly tuned from their wet chemical synthesis (Chapter 1 and Section 2.1) [57], the stability, reactivity, and biocompatibility of these structures largely depend on the coating of the metallic surface [58]. In the last two decades, several coating protocols, in particular for AuNSs, were described. In this section, the processes usually employed to stabilize AuNPs in physiological media and their modifications with (bio)molecules of interest are discussed [48, 59–64].

The general protocol to stabilize water-synthesized AuNPs relies on the exchange of the reaction coating with ligand molecules (ω-functionalized molecules) that usually comprise (i) a ligand group that is reactive toward the metal surface, and (ii) a hydrophilic group on the other end, that is, the ω-side [65]. Appropriate ligand groups must be chosen for each metal. For example, thiol groups can make pseudocovalent bonds (about 45 kcal/mol) with gold, while amines are the best coordinator groups for silver nanostructures [61, 66]. The general structure of the organic ω-functionalized molecules employed for the coating of AuNPs consists of a carbon chain with a thiol at one end and a carboxylic acid at the other one, such as mercaptoundecanoic acid or thiol derivative polyethylene glycol–COOH.

The stability of the native colloids relies on the electrostatic repulsion between the charged NPs, which are usually covered by negatively charged mother surfactants. The desorption of the mother surfactants upon chemisorptions of thiols may decrease the electrostatic stability of AuNPs during the ligand exchange on the metal surface and sometimes cause their irreversible aggregations caused by van der Waals interactions, whose impact increases with the diameter of the nanostructures. Also, if the hydrophilic end group is positive (as in the case of amines at pH ≤ 10), electrostatic interactions with negative desorbed surfactants may cause irreversible precipitation of the colloid [67]. As suggested by Ojea-Jimenez and Púntes [67], this may explain the intrinsic instability of citrate-synthesized AuNSs (Chapter 1), where desorbed citrate anions can form a bridge between amine-modified NPs. In the last decade, several methods were developed to improve the surface substitution on gold nanostructures. These methods are usually based on a multistep strategy that ensures the assembly of the thiol protection layer by slowing down the surfactant exchange [59, 68]. An example is shown in Figure 2.10. The first step consists of the displacement of the mother ligands, such as citrate, by a relatively short and negatively charged dithiol such as thioctic acid (TA). In the second step, a different bifunctionalized thiol ligand is added in large excess in order to displace the TA.

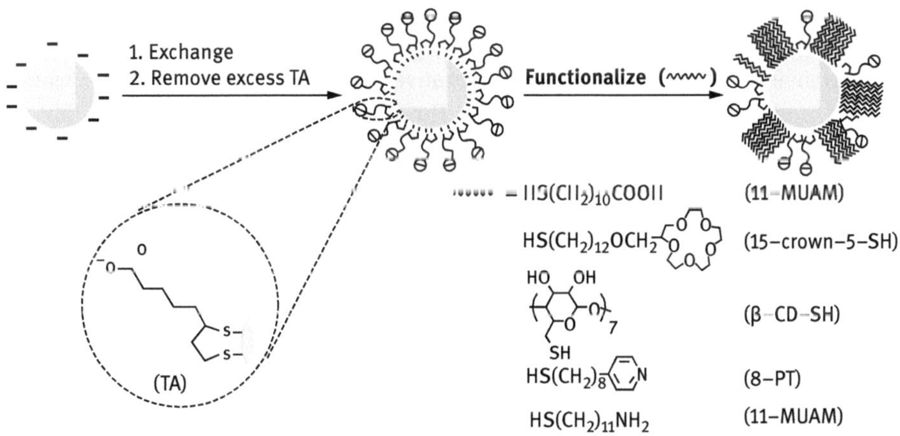

Figure 2.10: Schematic illustration of the coating process proposed by Lin and coworkers. In the first step, AuNSs are coated by thioctic acid (TA). In the second step, another thiol ligand with the desired ω-functionalization is added, and replaces TA. 11-MUDA, 11-mercaptoundecanoic acid; 15-crown-5-SH, 2-(12-mercaptododecyloxy)methyl-15-crown-5 ether; β-CD-SH, 6-mercapto -β-cyclodextrin; 8-PT, 8-(4-pyridyl)octanethiol; 11-MUAM, 11-mercaptoundecylamine. Reproduced with permission from S.Y. Lin, Y.T. Tsai, C.C. Chen, C.M. Lin and C.H. Chen, Journal of Physical Chemistry B, 2004, 108, 2134. ©2004, American Chemical Society [59].

This approach provides sufficient time to composite a densely packed monolayer of the second surfactant because TA has a slow displacement rate. By employing this protocol, stable AuNSs exhibiting negatively charged (carboxylate) or neutral [crown ethers or cyclodextrin (CD)] or positively charged (pyridinium or ammonium) moieties are achieved, depending on the ω-group of the second thiol (Figure 2.10).

The method previously described produces stable monofunctionalized AuNSs with a complex and low-rate process. Another widely used approach to stabilize and functionalize AuNSs is based on ω-functionalized thiol ligands comprising thiol–PEG derivatives (X-PEG-SH) [69, 70]. These polymers link to the metal surface by the thiol group and stabilize the nanostructures due to their high solubility in water, which arises from the ether structure of the chain. The reactive groups on the surface of encapsulated nanostructures can be easily varied by changing the end group of the polymers or by mixing different X-PEG-SH molecules as, for example, a 1:1 molar ratio of ω-amine- and ω-carboxy-PEG-SH [70]. The biosafety of AuNSs encapsulated by this method is usually acceptable [71, 72]. Usually, PEG polymers employed for the encapsulation have long chains of 3–5,000 Da and this may lead to an enlargement of 5–15 nm in the NP diameter. Furthermore, the folding of the X-PEG-SH chain on the metallic surface can trap the X functional groups, decreasing the total reactivity of the colloids or the exposure of the covalently linked (bio)molecules, that is, varying the chemical and physical behaviors.

A particularly interesting strategy to stabilize/functionalize water-synthesized AuNPs is based on peptides [73]. To encapsulate AuNPs, the group of Lévy and coworkers used *N*-cysteine peptides composed of three to five amino acids. The ones with a C-terminal carboxylic acid, that is, negatively charged, result in stable, water-soluble, and monofunctionalized colloids. However, the colloids became unstable if other functional groups such as amines were inserted in the last amino acid. By a careful design of the peptides sequence, multifunctionalized AuNSs stabilized due to the formation of a self-assembled monolayer (SAM) were obtained [61]. It is useful to remember that a SAM is formed when molecules are organized in a well-packed layer on a surface, making the disruption of the coating energetically unfavorable. This effect improves dramatically the stability of the encapsulated AuNPs.

Other strategies consist of, for example, the addition of a second ligand layer without exchanging the original one (Figure 2.11).

This approach is mainly employed for nanostructures synthesized in organic media and, thus, coated with hydrophobic surfactant. The second layer consists of amphiphilic molecules that can intercalate into the first hydrophobic surfactant layer with the hydrophobic portion and ensure water solubility of the nanocrystal because of the hydrophilic groups in their structure (Figure 2.11). Interlocked bilayers are formed, for example, by the addition of *n*-alkanoic acids or cetyltrimethylammonium bromide on AuNSs synthesized by the *Brust method* (Chapter 1) [74, 75]. Alternatively, also CD can be used for the formation of the second ligand layer. The hydrophobic cavity of the CD is penetrated by the hydrophobic tails of the first layer and its hydrophilic surface points toward the solution [76]. In these examples, the second layer is solely stabilized around the first layer by hydrophobic interactions. Pellegrino and coworkers

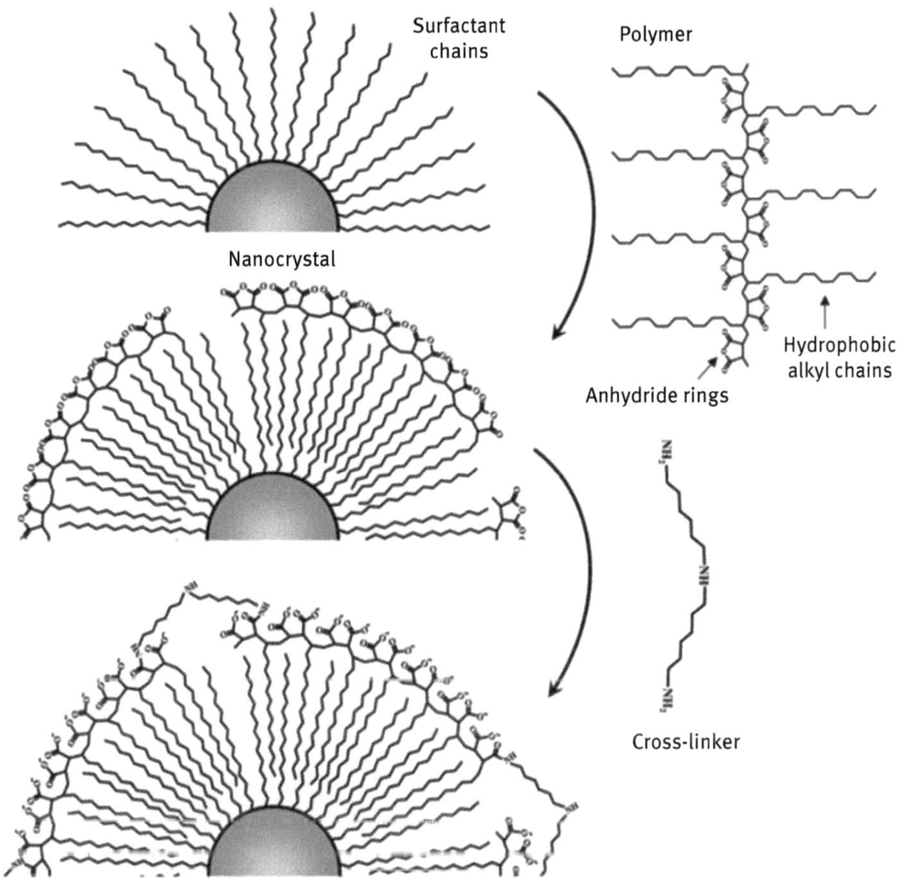

Figure 2.11: Polymer coating of hydrophobic nanoparticles proposed by Pellegrino and coworkers. The hydrophobic alkyl chains of the polymer intercalate into the surfactant coating. The anhydride rings are localized on the surface of the polymer-coated nanocrystal. The amino end groups of the cross-linker molecule open the rings and link the individual polymer chains. The surface of the polymer shell becomes negatively charged, stabilizing the particles in water by electrostatic repulsion. Reproduced with permission from T. Pellegrino, L. Manna, S. Kudera, T. Liedl, D. Koktysh, A.L. Rogach, S. Keller, J. Radler, G. Natile and W.J. Parak, Nano Letters, 2004, 4, 703. ©2004, American Chemical Society [60].

suggested an approach in which an amphiphilic polymer, [poly(maleic anhydride-alt-1-tetradecene)], is employed in order to, after intercalation in the mother surfactant layer of AuNPs, further stabilize the new polymeric layer by cross-linking through a biamine linker (Figure 2.11) [60]. It is worth to remember that these strategies are time-consuming, and their employment require skilled operators.

In conclusion, the right coating is a pivotal subject to consider for AuNPs in order to support their potentiality in nanomedicine. Indeed, besides stability and potential modification with (bio)molecules, the interactions of NPs with living cells or organisms are strictly correlated to the surface behaviors of NPs.

2.3 Biological features

In agreement to many reports, AuNPs are internalized by cells and seem low toxic [66, 77–79]. On the other hand, investigations that focus on biodistribution and biosafety of AuNPs are relatively few and sometimes controversial [77]. Conflicting results may arise from several variables, such as the variability of the toxicity assays, cell lines, AuNPs size/geometry dispersion, and coatings. This topic is fully addressed elsewhere [80]. In this section, the prominent results on the processes of internalization and on the toxicity of AuNPs on living cells and for organisms are reported. For a more comprehensive discussion of this topic, the readers are encouraged to refer to more specialized works [81].

The plasma membrane is a selectively permeable lipid bilayer that defines the boundary and maintains the essential intracellular environment of the cell. Small and nonpolar molecules such as oxygen and carbon dioxide can readily diffuse across the membrane; however, nanomaterials or polar molecules such as ions are generally unable to cross the plasma membrane on their own. In nature, important ions and nanometer-sized proteins can be transported across the lipid bilayer through specialized membrane transport protein channels [82]. Most other nanoscale macromolecules and molecular assemblies are internalized through the endocytic pathway upon contact with the cell membrane (Figure 2.12). In this process, macromolecules are internalized in cells by their enclosure in membrane vesicles.

By following this pathway, nanomaterials are usually confined in endosomes and not able to reach the cytosol because trapped in endolysosomal vesicles unless cointernalized with membrane-disrupting agents [84]. On this hand, several approaches have been proposed to transport nanomaterials directly into the cytosol of cells, such as (i) disruption of endosomes by the *proton sponge effect* mechanism (typical for polycations that have a buffering capacity below the physiological pH) or the use of chloroquine [85, 86], (ii) direct microinjection of nanomaterials into cells [87], (iii) use of electroporation [88], and (iv) conjugation of natural cell-penetrating peptides to nanomaterials [89].

Thus, in the absence of internalization agents linked to the nanostructures, AuNPs usually penetrate cells by a two-step process that involves their adsorption onto the cell surface followed by endocytosis (Figure 2.12). Consequently, living cell uptake of AuNPs may be strongly influenced by (i) AuNPs geometry (size and shape) and (ii) AuNPs charge.

The uptake of 14, 50, and 74 nm AuNSs was investigated in HeLa cells, as shown in Figure 2.13 [90].

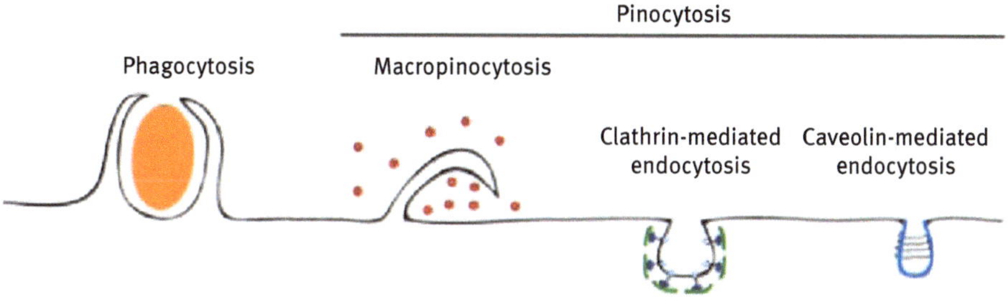

Figure 2.12: Major pathways for endocytosis, the process by which cells internalize external objects. By phagocytosis, cells can internalize solid matter larger than 0.75 µm, while by pinocytosis the cells can take up external solutions through 50–1,000 nm cell membrane invagination (0.5–1 µm for macropinocytosis, about 100 nm for clathrin-mediated, and about 50 nm for caveolae-mediated endocytosis). It is important to note that phagocytosis and macropinocytosis occur in a nonspecific manner, while clathrin- and caveolae-mediated endocytosis happen specifically by membrane receptor-mediated pathways. Reproduced with permission from A. Verma and F. Stellacci, Small, 2010, 1, 12. ©2010, Wiley-VCH [83].

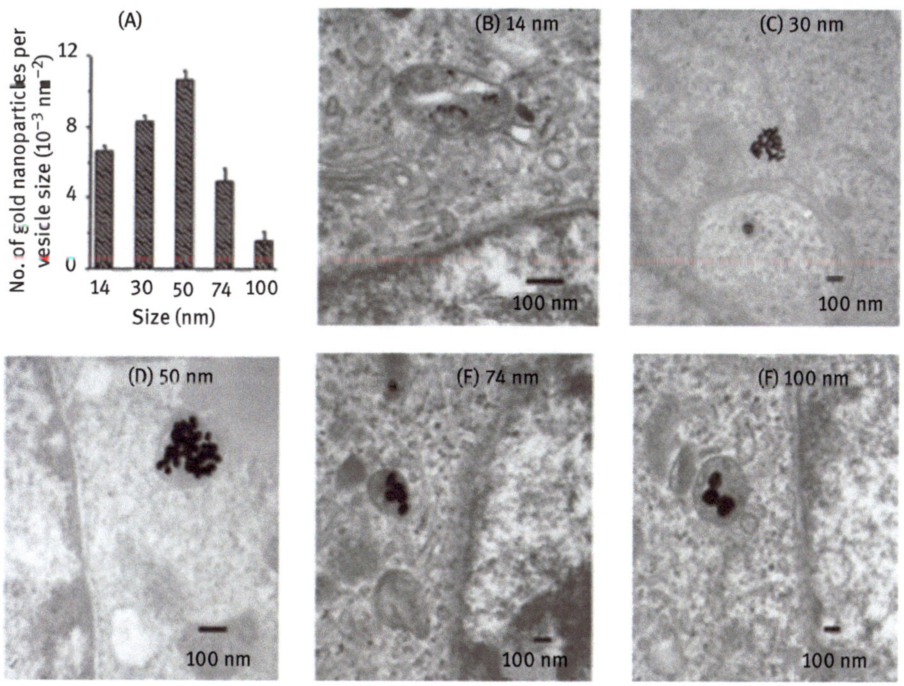

Figure 2.13: TEM images of gold nanoparticles entrapped in vesicles within HeLa cells. (a) Graph showing the number of gold nanoparticles per vesicle diameter versus nanoparticle size. TEM images of nanoparticles with a diameter of (b) 14 nm, (c) 30 nm, (d) 50 nm, (e) 74 nm, and (f) 100 nm within vesicles. Reproduced with permission from B.D. Chithrani, A.A. Ghazani and W.C. Chan, Nano Letters, 2006, 6, 662. ©2006, American Chemical Society [90].

The kinetics of uptake as well as the saturation concentration of AuNSs varied with the NP diameter and the main internalization was achieved with about 50 nm spheres. The authors suggest that there might be an optimal size for efficient nanomaterial uptake in cells. This investigation also concludes that AuNPs are benign and biologically inert in the size range of 10–100 nm. When the gold core size is reduced to below 2 nm, the surface of the AuNPs may show unusual chemical reactivity. For example, some AuNPs in the cluster regime, that is, with a diameter less than 2 nm, may act as efficient catalysts due to the high surface reactivity [91]. Moreover, AuNPs with a diameter of 1.4 nm showed significant toxicity to culture cells via induction of oxidative stress and mitochondrial damage [91]. On the other hand, it is worth to report that some very promising nanoarchitectures for medical applications comprise ultrasmall NPs and have demonstrated an impressive biosafety profile in cultured cells, zebrafish, and murine models [30, 66, 79, 92]. Indeed, ultrasmall NPs may exhibit cytotoxic profile when the stabilizing ligands allow for direct access to the metal surface either for the direct interaction with biomolecules or for catalytic activity of the unshielded metal surface [29]. Remarkably, the inherent polydispersity of batch of NPs may cause unpredictable events and different cytotoxicity profiles [80]. Accordingly, the control on the monodispersity of the geometry of the nanostructures is a pivotal variable to consider in order to develop effective nanoprobes (Chapter 1).

Another widely investigated factor affecting the rate of internalization of nanomaterials is their shape [80]. In general, spherical particles are internalized up to five times more rapidly than rod-shaped particles of a similar size [93]. This result is explained by the longer membrane wrapping time required for the elongated particles. On this hand, also gold nanocubes are more difficulty internalized than gold NSs, probably because of the initial higher surface contact area between the nanostructures and the cell membrane [94]. Data on the dependence of cellular uptake on the shape of nanomaterials have to be usually taken with care, as the comparison among NPs with noncomparable sizes or with different surface functionalization are reported. Hence, highlight a net shape dependence whereas other physiochemical parameters are not completely fixed can be a difficult task [81].

The dependence of living cells uptake on AuNPs geometry is strongly influenced by the surface coating [94].

Indeed, the functional groups on the AuNPs surface dramatically affect several properties, such as colloidal stability, solubility, and interactions with (bio)molecules or cell membrane (Figure 2.14 and Section 2.2). For example, while neutral functional groups such as ethers or azides usually prevent unwanted interactions between nanomaterials and biological matter (such as proteins), charged functional groups, such as amines and carboxylic acid, are responsible for active interactions, also with cells. On this regard, many efforts have been profuse to the design of agents, among which peptides, able to modulate the interactions between nanomaterials and (bio)molecules, that is, the protein corona, and consequently their interactions with living matter [95, 96].

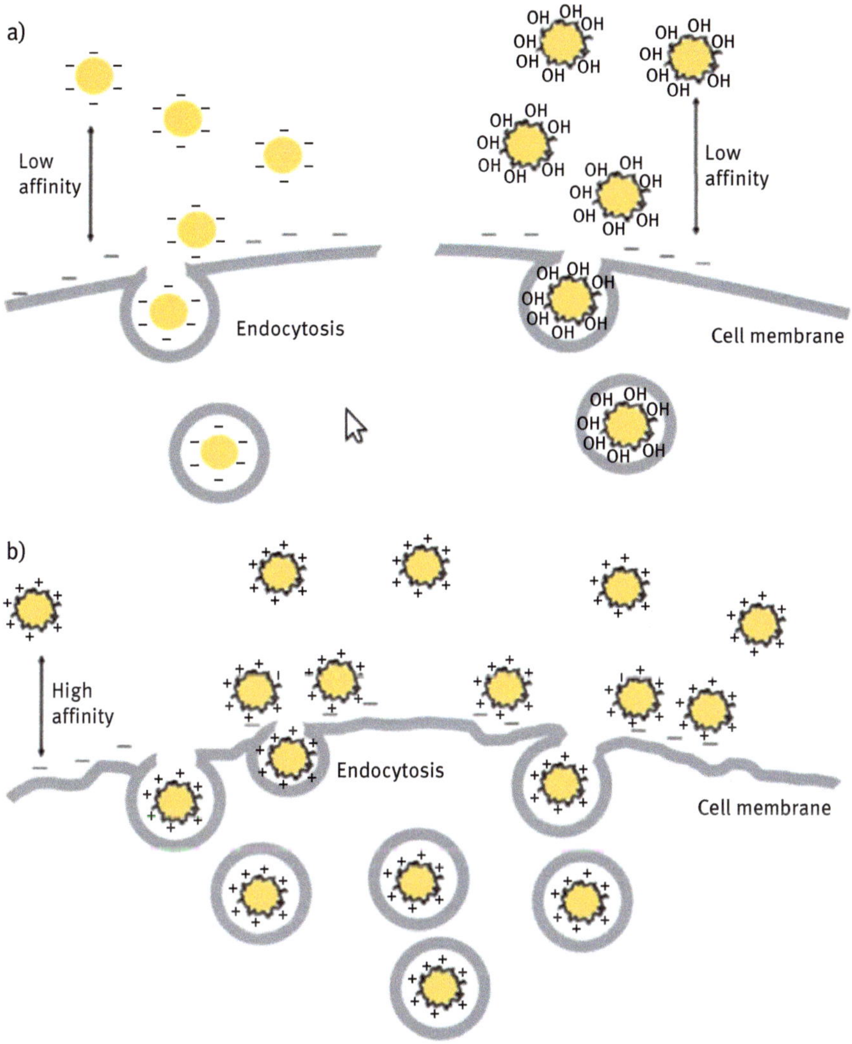

Figure 2.14: Schematic illustration of the endocytosis pathways for AuNP featuring different surface charges. (a) Citrate-coated (negative charge) and polyvinyl alcohol (PVA)-coated (neutral charge) NP display low affinity of interaction with the cell membrane, while (b) a poly(allylamine hydrochloride)-coated NP (positive charge) shows high cell-membrane binding affinity. Reproduced with permission from A. Verma and F. Stellacci, Small, 2010, 1, 12. ©2010, Wiley-VCH [83].

In general, neutral and negatively charged NPs are less adsorbed on the negatively charged cell membrane surface, resulting in a lower level of internalization as compared to the positively charged particles, demonstrating the role of surface charge in the internalization of NP [97, 98]. However, there are several evidences of the uptake of negatively charged particles despite the unfavorable interaction between the particles and the negatively charged cell membrane. Besides other processes, the internalization of negatively charged NPs may occur through nonspecific binding (and sometimes clustering) of the particles on cationic sites on the plasma membrane followed by their endocytosis [83]. Another proposed mechanism relies on the adsorption of serum proteins on the surface of NP through electrostatic and hydrophobic interactions, which allowed the AuNPs to interface with the cell membrane (Figure 2.15) [99].

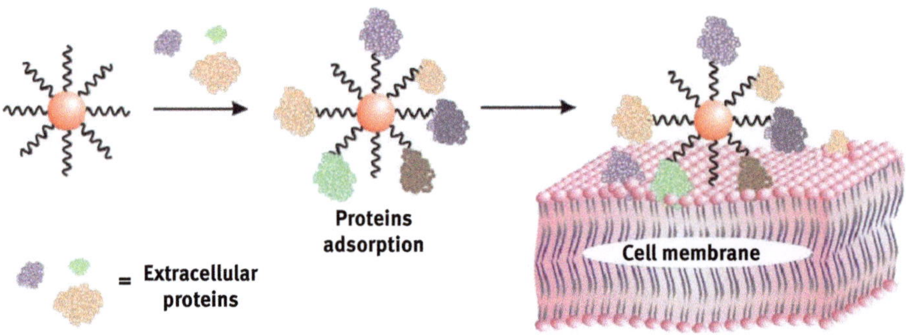

Figure 2.15: Schematic view of the interactions among negatively charged coated NPs, proteins, and cell membranes. Proteins in the media may adsorb on negative AuNPs, allowing to their interface with the cell membrane and the subsequent internalization. Reproduced with permission from A. Verma and F. Stellacci, Small, 2010, 1, 12. ©2010, Wiley-VCH [83].

In general, positive charged NPs are more cytotoxic than their negative counterpart [100]. The cause of this behavior seems to depend on the effectiveness of cationic AuNPs to cross cell membrane barriers and localize inside the compartment of the cell [82]. However, data regarding the dependence of toxicity by the charge are affected by an important factor: the absorption on the surface of AuNPs of cellular proteins present in the medium. For example, cationic gold nanorods that show a zeta potential of +40 mV become anionic upon mixing with cell growth media containing 10% bovine serum albumin [77]. Indeed, serum albumin adsorbs to the surface of cationic AuNPs and causes the binding to the surface of the cell as a single anionic complex. Albumin binds to four types of cellular surface receptors that induce endocytosis. Thus, albumin can act as an endocytotic ligand for nonfunctionalized AuNPs. Interestingly, the capping agents (or the building blocks resulting from their degradation) used to stabilize the AuNPs have to be fully considered for cytotoxicity assessment. For example, the toxicity recorded from gold nanorods is mainly caused by the capping agent (usually the quaternary amine hexadecyltrimethylammonium bromide, CTAB) than from the nanostructure itself [101].

In conclusion, cellular uptake and cytotoxicity of AuNPs depend on many variables, such as size, shape, surface functionality/charge, aggregation and concentration of AuNPs, cell line, incubation conditions, viability assays, and type of culture media. Nowadays, there is a lack of standardization for the evaluation of the interactions between nanomaterials and biological matter [102].

In the following section, a short overview on the techniques generally employed (but not standardized) to observe the presence and the localization of AuNPs in the various compartments of the cell is presented. The cellular uptake of AuNPs in living cells can be qualitatively measured by imaging techniques such as TEM or confocal microscopy. TEM on fixed cells is a standard method to determine AuNPs in cellular compartments providing structural details at nanometer resolution [92]. Moreover, AuNPs are electron dense and can be observed quite easily by TEM. TEM analyses suffer from some drawbacks such as (i) the need to prepare thin specimen samples (50–200 nm in thickness), (ii) long processing time to prepare specimens for imaging, (iii) some characteristics of cells and particles can be missed upon sectioning, and (iv) very long analysis time to have statistically valid results. On the other hand, confocal microscopy can provide real-time results on living cells even if the conjugation of AuNPs to dyes is usually required. Interestingly, the only technique for the quantitative detection of AuNPs in biological matter is the inductively coupled plasma mass spectrometry (ICP-MS). ICP-MS enables the fine quantification of gold in biological specimens (down to few ppt) but this type of analysis requires the degradation of the samples, is destructive, and is expensive.

Even though 2D and 3D living cell models are a simple, fast, and cheap method to provide mechanistic and molecular understanding on the interactions between AuNPs and cellular components, they cannot be used to extrapolate speculations on the in vivo fate of AuNPs [103, 104]. Obviously, a whole organism is much more complex than a single cell and, after administration, nanomaterials encounter a series of hurdles in their path toward the target for which they have been designed. Therefore, in order to translate AuNPs to clinical practice, their biosafety and biokinetics (absorption, biodistribution, and fate) must be evaluated in a couple of models, among which zebrafish, mice, and rabbits [66]. Interestingly, while the in vivo efficacy of nanomaterials for healthcare has been widely investigated, their biokinetics (the absorption, distribution, metabolism, and elimination – for nanomaterials), although essential for translation as well as the efficacy, have only rarely been evaluated [105]. Indeed, up to now the main focus of the field was to describe and demonstrate therapeutic and diagnostic efficacy of NPs [106]. According to the nanomedicine 2.0 model, and also considering the "3Rs concept," a broad comprehension of metal biokinetics is pivotal [30, 106, 107]. In general, AuNPs can be administered by all the routes generally used for drugs, such as inhalation, oral, intestinal, and intravenous [105]. In general, AuNPs accumulate in all the excretory organs with a strong dependence on their size and a weak one on their charge, probably for absorption on the surface of the AuNPs of plasma proteins [108, 109]. AuNPs with a diameter >10 nm usually accumulate in the liver, kidney, and spleen, while the ones with diameter <10 nm can also accumulate in the brain, heart, lung, and blood (Figure 2.16) [108–110].

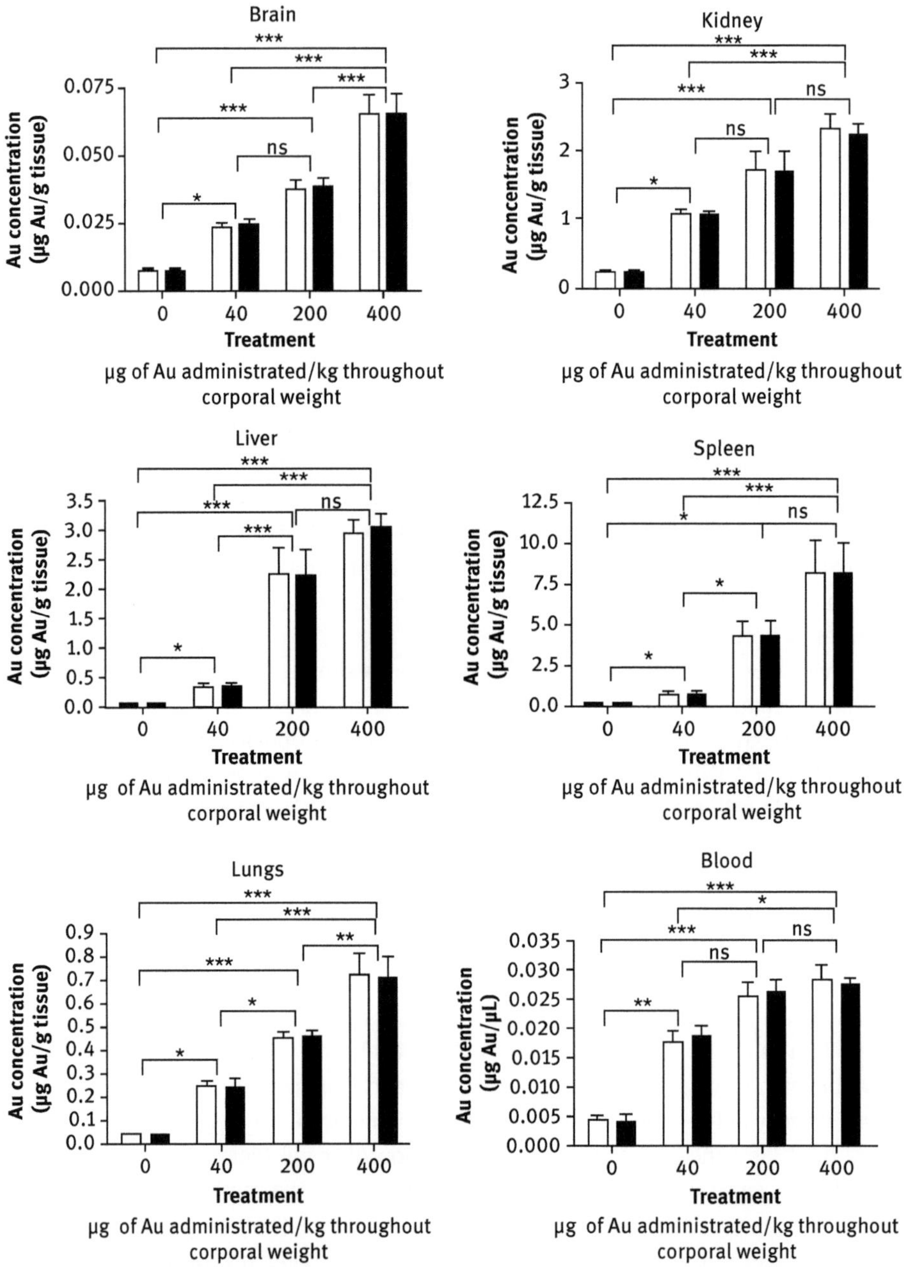

Figure 2.16: Gold accumulation in body tissues. Animals were injected intraperitoneally with various concentrations of citrate-capped AuNP with a size of 12.5 nm (40, 200, and 400 µg/kg) for eight consecutive days. After one day of withdrawal following the last injection, animals were sacrificed and blood and organs were extracted and analyzed by graphite furnace – atomic absorption spectrometry (GF-AAS) (white bar) and ICP-MS (black bar) to measure accumulated gold.

It is worth to remember that biodistribution behaviors of nanomaterials are strictly dependent on their hydrodynamic diameter rather than their physical size [30]. Moreover, the pharmacokinetic parameters of AuNPs may vary when peculiar molecules such as PEG or antibodies are present on their surface. For example, Niidome and coworkers have observed that the blood circulation half-time of gold nanorods can be increased by displacing the CTAB with PEG [111]. This effect can be correlated to the reduction of nonspecific adsorption of plasma proteins onto the surfaces of gold nanorods due to the presence of PEG.

If a neoplasm is present in the model, AuNPs can also passively accumulate there by enhanced permeability and retention (EPR) effect (Figure 2.17) [112–114].

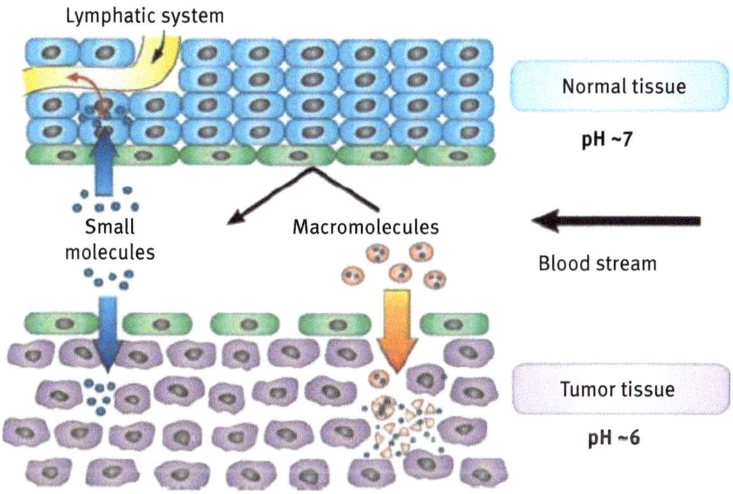

Figure 2.17: A comparison between normal (upper) and tumor tissues. In the first, the cells are well organized and compacted, and a functional lymphatic system is also present. The pH of the tissues is different, indeed the tumor one has a more acidic pH. Reproduced with permission from S. Bisht and A. Maitra, Wiley International Review: Nanomedicine and Nanobiology, 2009, 1, 415. ©2009, Wiley-VCH [112].

EPR effect is caused by the poor organization of the tissues of some neoplasms together with a local more acid environment and a poor lymphatic system [115]. Even if the EPR effect is controversial, several evidences confirmed that it enhances the passive accumulation of nanomaterials in tumors, increasing their specificity of action [116].

Figure 2.16 (continued)
The values correspond to the gold concentration relative to the weight of the dry tissue. Bars represent mean ± standard error. Reproduced with permission from C. Lasagna-Reeves, D. Gonzalez-Romero, M.A. Barria, I. Olmedo, A. Clos, V.M.S. Ramanujam, A. Urayama, L. Vergara, M.J. Kogan and C. Soto, Biochemical and Biophysical Research Communications, 2010, 393, 649. ©2010, Elsevier [110].

It is important to remember that AuNPs do not biodegrade, and if the hydrodynamic size is over than the size of glomeruli in kidneys (<6 nm), they cannot cross the filtration barriers and persist in organisms for indefinite time [28]. On the other hand, therapeutics have not to leave residues in patients after the action, as suggested by the approbation agencies [29, 117]. This hurdle has prevented the translation of AuNPs to the clinical practice in the last 20 years. Very recently, the ultrasmall-in-nano approach has been proposed to design promising nanoarchitectures that are able to conjugate the most intriguing features of AuNPs with their excretion from organism [20, 26].

Far from comprehensively discussing this complex and interesting topic, the readers are encouraged to further explore the subject in more focused works [28–30, 66, 81].

References

[1] Burda C, Chen XB, Narayanan R, El-Sayed MA. Chemistry and properties of nanocrystals of different shapes. Chem Rev 2005;105(4):1025–1102.

[2] Kelly KL, Coronado E, Zhao LL, Schatz GC. The optical properties of metal nanoparticles: the influence of size, shape, and dielectric environment. J Phys Chem B 2003;107(3):668–677.

[3] Faraday M. Experimental relations of gold (and other metals) to light. Philos Trans R Soc London 1857;147:145.

[4] Krahne R, Morello G, Figuerola A, George C, Deka S, Manna L. Physical properties of elongated inorganic nanoparticles. Phys Rep 2011;501(3–5):75–221.

[5] Creighton JA, Eadon DG. Ultraviolet-visible absorption spectra of the colloidal metallic elements. J Chem Soc Faraday Trans 1991.

[6] Njoki P, Lim I, Mott D, Park H. Size correlation of optical and spectroscopic properties for gold nanoparticles. J Phys Chem C 2007;(111):14664–14669.

[7] Ashcroft NW, Mermin ND. Solid State Physics 1976. ISBN: 0030839939, 9780030839931.

[8] Link S, El-Sayed M. Optical properties and ultrafast dynamics of metallic nanocrystals. Annu Rev Phys Chem 2003;54:331–366.

[9] Rosi NL, Mirkin CA. Nanostructures in biodiagnostics. Chem Rev 2005;105(4):1547–1562.

[10] González AL, Reyes-Esqueda JA, Noguez C, Gonzalez A. Optical properties of elongated noble metal nanoparticles nanoparticles. J Phys Chem C 2008;112(19):7356–7362.

[11] Link S, El-Sayed M. Shape and size dependence of radiative, non-radiative and photothermal properties of gold nanocrystals. Int Rev Phys Chem 2000;19(3):409–453.

[12] Mie G. Grundlagen einer theorie der materie. Ann Phys 1912;11:1.

[13] Lim I, Zhong C. Molecularly-mediated assembly of gold nanoparticles. Gold Bull 2007;40(1): 59–66.

[14] Jain PK, Lee KS, El-Sayed IH, El-Sayed MA. Calculated absorption and scattering properties of gold nanoparticles of different size, shape, and composition: applications in biological imaging and biomedicine. J Phys Chem B 2006;110(14):7238–7248.

[15] Sonnichsen C, Reinhard BM, Liphardt J, et al. A molecular ruler based on plasmon coupling of single gold and silver nanoparticles. Nat Biotechnol 2005;23(6):741–745.

[16] Sokolov K, Follen M, Aaron J, Pavlova I. Real-time vital optical imaging of precancer using anti-epidermal growth factor receptor antibodies conjugated to gold nanoparticles. Cancer Res 2003;63(9):1999–2004.
[17] Jain P, Huang X. Noble metals on the nanoscale: optical and photothermal properties and some applications in imaging, sensing, biology, and medicine. Accounts Chem ... 2008;41: 1578–1586.
[18] Avigo C, Cassano D, Kusmic C, Voliani V, Menichetti L. Enhanced photoacoustic signal of passion fruit-like nanoarchitectures in a biological environment. J Phys Chem C 2017;121(12): 6955–6961.
[19] Armanetti P, Pocoví-Martínez S, Flori A, et al. Dual photoacoustic/ultrasound multi-parametric imaging from passion fruit-like nano-architectures. Nanomedicine Nanotechnology, Biol Med 2018;14(6):1787–1795.
[20] Cassano D, Santi M, D'Autilia F, Mapanao AK, Luin S, Voliani V. Photothermal effect by NIR-responsive excretable ultrasmall-in-nano architectures. Mater Horizons 2019;6 (3):531–537.
[21] Huang X, Neretina S, El-Sayed MA. Gold nanorods: from synthesis and properties to biological and biomedical applications. Adv Mater 2009;21(48):4880–4910.
[22] Gans R. Über die Form ultramikroskopischer Silberteilchen. Ann Phys 1915;352(10):270–284.
[23] Papavassiliou GC. Optical properties of small inorganic and organic metal particles. Prog Solid State Chem 1979;12(3–4):185–271.
[24] Sau TK, Murphy CJ. Room temperature, high-yield synthesis of multiple shapes of gold nanoparticles in aqueous solution. J Am Chem Soc 2004;126(28):8648–8649.
[25] Eichler J, Knof J, Lenz H. Measurements on the depth of penetration of light (0.35–1.0 um) in tissue. Radiat Environ Biophys 1977;14(3):239–242.
[26] Cassano D, Rota Martir D, Signore G, Piazza V, Voliani V. Biodegradable hollow silica nanospheres containing gold nanoparticle arrays. Chem Commun 2015;51(49):9939–9941.
[27] Cassano D, David J, Luin S, Voliani V. Passion fruit-like nano-architectures: a general synthesis route. Sci Rep 2017;7:43795.
[28] Cassano D, Pocoví-Martínez S, Voliani V. Ultrasmall-in-nano approach: enabling the translation of metal nanomaterials to clinics. Bioconjug Chem 2018;29(1):4–16.
[29] Vlamidis Y, Voliani V. Bringing again noble metal nanoparticles to the forefront of cancer therapy. Front Bioeng Biotechnol 2018;6:143.
[30] Cassano D, Summa M, Pocoví-Martínez S, et al. Biodegradable ultrasmall-in-nano gold architectures: mid-period in vivo distribution and excretion assessment. Part Part Syst Charact 2019;36(2):1800464.
[31] Storhoff JJ, Lazarides A a, Mucic RC, Mirkin C a, Letsinger RL, Schatz GC. What controls the optical properties of DNA-linked gold nanoparticle assemblies? J Am Chem Soc 2000;122(19): 4640–4650.
[32] Jain PK, Huang WY, El-Sayed M a, Visiting M, KJ P, Qian W. On the universal scaling behavior of the distance decay of plasmon coupling in metal nanoparticle pairs: a plasmon ruler equation. Nano Lett 2007;7(7):2080–2088.
[33] Wei Q-H-H, Su K-H-H, Durant S, Zhang X. Plasmon resonance of finite one-dimensional Au nanoparticle chains. Nano Lett 2004;4(6):1067–1071.
[34] Su K, Wei Q, Zhang X, Mock J. Interparticle coupling effects on plasmon resonances of nanogold particles. Nano Lett 2003;3(8):1087–1090.
[35] Reinhard BM, Siu M, Agarwal H, Alivisatos AP, Liphardt J. Calibration of dynamic molecular rulers based on plasmon coupling between gold nanoparticles. Nano Lett 2005;5(11): 2246–2252.

[36] Weiss S. Fluorescence spectroscopy of single biomolecules. Science (80-) 1999;283(5408): 1676–1683.
[37] Mooradian A. Photoluminescence of metals. Phys Rev Lett 1969;22:185.
[38] Mohamed M, Volkov V, Link S, El-Sayed M. Thelightning'gold nanorods: fluorescence enhancement of over a million compared to the gold metal. Chem Phys Lett 2000;317(6): 517–523.
[39] Wu X, Ming T, Wang X, Wang P, Wang J, Chen J. High-photoluminescence-yield gold nanocubes: for cell imaging and photothermal therapy. ACS Nano 2010;4(1):113–120.
[40] Wilcoxon J, Martin J. Photoluminescence from nanosize gold clusters. J Chem ... 1998;108 (21):9137–9143.
[41] Le Guével X, Hötzer B, Jung G, Hollemeyer K, Trouillet V, Schneider M. Formation of fluorescent metal (Au, Ag) nanoclusters capped in bovine serum albumin followed by fluorescence and spectroscopy. J Phys Chem A 2011;115(22):10955–10963.
[42] Yahia-Ammar A, Sierra D, Mérola F, Hildebrandt N, Le Guével X. Self-assembled gold nanoclusters for bright fluorescence imaging and enhanced drug delivery. ACS Nano 2016;10 (2):2591–2599.
[43] Noguez C. Surface plasmons on metal nanoparticles: the influence of shape and physical environment. J Phys Chem C 2007;111(10):3806–3819.
[44] Bohren C, Huffman D. Absorption and scattering of light by small particles, New York, Wiley, 2008.
[45] Kreibig U. Electronic properties of small silver particles: the optical constants and their temperature dependence. J Phys F Met Phys 1974;4(7):999–1014.
[46] Baida H, Billaud P, Marhaba S, et al. Quantitative determination of the size dependence of surface plasmon resonance damping in single Ag@SiO_2 nanoparticles. Nano Lett 2009;9(10): 3463–3469.
[47] Meier M, Wokaun A. Enhanced fields on large metal particles: dynamic depolarization. Opt Lett 1983;8(11):581–583.
[48] Voliani V, Ricci F, Signore G, Nifosì R, Luin S, Beltram F. Multiphoton molecular photorelease in click-chemistry-functionalized gold nanoparticles. Small 2011;7(23):3271–3275.
[49] Mohammadi A. Fluorescence enhancement with the optical (Bi-) conical antenna†. J Phys Chem A 2010;114(16):7372–7377.
[50] Albrecht M, Creighton J. Anomalously intense Raman spectra of pyridine at a silver electrode. J Am Chem ... 1977;99(15):5215–5217.
[51] Jeanmaire D, Van DR. Surface Raman spectroelectrochemistry: Part I. Heterocyclic, aromatic, and aliphatic amines adsorbed on the anodized silver electrode. J Electroanal Chem Interfacial Electrochem 1977;84(1):1–20.
[52] Moskovits M. Surface roughness and the enhanced intensity of Raman scattering by molecules adsorbed on metals. J Chem Phys 1978;69(9):4159–4161.
[53] Schatz G. Theoretical studies of surface enhanced Raman scattering. Acc Chem Res 1984;17 (10):370–376.
[54] Alvarez-Puebla R a, Liz-Marzán LM. SERS-based diagnosis and biodetection. Small 2010;6(5): 604–610.
[55] Wang Y, Yan B, Chen L. SERS tags: novel optical nanoprobes for bioanalysis. Chem Rev 2013;113(3):1391–1428.
[56] Guerrini L, Alvarez-Puebla R a. Surface-enhanced Raman spectroscopy in cancer diagnosis, prognosis, and monitoring. Cancers (Basel) 2019;11(6):748.
[57] Orendorff C, Sau T, Murphy C. Shape-dependent plasmon-resonant gold nanoparticles. Small 2006;5(2):636–639.

[58] Yin Y, Alivisatos AP. Colloidal nanocrystal synthesis and the organic-inorganic interface. Nature 2005;437(7059):664–670.
[59] Lin S-Y-Y, Tsai Y-T-T, Chen -C-C-CH, Lin C-M-M. Two-step functionalization of neutral and positively charged thiols onto citrate-stabilized au nanoparticles. J Phys Chem B 2004;108 (7):2134–2139.
[60] Pellegrino T, Manna L, Kudera S, et al. Hydrophobic nanocrystals coated with an amphiphilic polymer shell: A general route to water soluble nanocrystals. Nano Lett 2004;4(4):703–707.
[61] Voliani V, Luin S, Ricci F, Beltram F. Single-step bifunctional coating for selectively conjugable nanoparticles. Nanoscale 2010;2(12):2783–2789.
[62] Gong SQ, Aryal S, Grailer JJ, Pilla S, Steeber DA. Doxorubicin conjugated gold nanoparticles as water-soluble and pH-responsive anticancer drug nanocarriers. J Mater Chem 2009;19(42): 7879–7884.
[63] Voliani V, Signore G, Vittorio O, et al. Cancer phototherapy in living cells by multiphoton release of doxorubicin from gold nanospheres. J Mater Chem B 2013;1(34):4225.
[64] Voliani V, Ricci F, Luin S, Beltram F. Peptidic coating for gold nanospheres multifunctionalizable with photostable and photolabile moieties. J Mater Chem 2012;22 (29):14487.
[65] Gittins D, Caruso F. Spontaneous phase transfer of nanoparticulate metals from organic to aqueous media. Angew Chemie Int Ed 2001;40(16):3001–3004.
[66] Cassano D, Mapanao AK, Summa M, et al. Biosafety and biokinetics of noble metals: the impact of their chemical nature. ACS Appl Biomater 2019.
[67] Ojea-Jiménez I, Puntes V, Ojea-Jimenez I. Instability of cationic gold nanoparticle bioconjugates: the role of citrate ions. J Am Chem Soc 2009;131(37):13320–13327.
[68] Aslan K, Pérez-Luna V. Surface modification of colloidal gold by chemisorption of alkanethiols in the presence of a nonionic surfactant. Langmuir 2002;(20):6059–6065.
[69] Perrault SD, Chan WCW. Synthesis and surface modification of highly monodispersed, spherical gold nanoparticles of: 50–200 nm. J Am Chem Soc 2009;131(47):17042–17043.
[70] Maus L, Spatz JP, Fiammengo R. Quantification and reactivity of functional groups in the ligand shell of PEGylated gold nanoparticles via a fluorescence-based assay. Langmuir 2009;25(14):7910–7917.
[71] Mahmoudi M, Sant S, Wang B. Superparamagnetic iron oxide nanoparticles (SPIONs): development, surface modification and applications in chemotherapy. Adv Drug Deliv Rev 2011;63:24–46.
[72] Lipka J, Semmler-Behnke M, Sperling R. Biodistribution of PEG-modified gold nanoparticles following intratracheal instillation and intravenous injection. Biomaterials 2010;31(25): 6574–6581.
[73] Lévy R, Thanh NTK, Doty RC, et al. Rational and combinatorial design of peptide capping ligands for gold nanoparticles. J Am Chem Soc 2004;126(32):10076–10084.
[74] Shen L, Laibinis P, Hatton T. Bilayer surfactant stabilized magnetic fluids: synthesis and interactions at interfaces. Langmuir 1999;15(2):447–453.
[75] Swami A, Kumar A, Sastry M. Formation of water-dispersible gold nanoparticles using a technique based on surface-bound interdigitated bilayers. Langmuir 2003;19(4):1168–1172.
[76] Wang Y, Wong J, Teng X, Lin X, Yang H. “Pulling” nanoparticles into water: phase transfer of oleic acid stabilized monodisperse nanoparticles into aqueous solutions of α-cyclodextrin. Nano Lett 2003;3(11):1555–1559.
[77] Alkilany AM, Murphy CJ. Toxicity and cellular uptake of gold nanoparticles: what we have learned so far?. J Nanoparticle Res 2010;12(7):2313–2333.
[78] Lewinski N, Colvin V, Drezek R. Cytotoxicity of nanoparticles. Small 2008;4(1):26–49.

[79] d'Amora M, Cassano D, Pocoví-Martínez S, Giordani S, Voliani V. Biodistribution and biocompatibility of passion fruit-like nano-architectures in zebrafish. Nanotoxicology 2018:1–9.
[80] Faria M, Björnmalm M, Thurecht KJ, et al. Minimum information reporting in bio–nano experimental literature. Nat Nanotechnol 2018;13(9):777–785.
[81] Cassano D, Voliani V. Behaviors and Persistence of Nanomaterials in Biomedical Applications, Wiley-Scrivener, 2018.
[82] Alberts B, Johnson A, Lewis J, Morgan D, Raff M, Roberts K, Walter P. Molecular biology of the cell (4th Ed). J Biol Educ 2002;37(1):45–47. ISBN: 9780815344322.
[83] Verma A, Stellacci F. Effect of surface properties on nanoparticle–cell interactions. Small 2010;1(6):12–21.
[84] Conner S, Schmid S. Regulated portals of entry into the cell. Nature 2003;422(6927):37–44.
[85] Hu Y, Litwin T, Nagaraja A, Kwong B. Cytosolic delivery of membrane-impermeable molecules in dendritic cells using pH-responsive core-shell nanoparticles. Nano ... 2007;7(10): 3056–3064.
[86] Sandhu K, McIntosh C. Gold nanoparticle-mediated transfection of mammalian cells. Bioconjugate Chem 2002;13(1):3–6.
[87] Wolff J, Malone R, Williams P, Chong W. Direct gene transfer into mouse muscle in vivo. Science (80-) 1990;247(4949):1465–1468.
[88] Rojas-Chapana J, Correa-Duarte M. Enhanced introduction of gold nanoparticles into vital Acidothiobacillus ferrooxidans by carbon nanotube-based microwave electroporation. Nano Lett 2004;4(5):985–988.
[89] Tkachenko A, Xie H. Multifunctional gold nanoparticle-peptide complexes for nuclear targeting. J Am Chem Soc 2003;125(16):4700–4701.
[90] Chithrani BD, Ghazani AA, Chan WCW. Determining the size and shape dependence of gold nanoparticle uptake into mammalian cells. Nano Lett 2006;6(4):662–668.
[91] Dreaden EC, Alkilany AM, Huang X, Murphy CJ, El-Sayed MA. The golden age: gold nanoparticles for biomedicine. Chem Soc Rev 2012;41(7):2740–2779.
[92] Cassano D, Santi M, Cappello V, Luin S, Signore G, Voliani V. Biodegradable Passion Fruit-Like Nano-Architectures as Carriers for Cisplatin Prodrug. Part Part Syst Charact 2016;33(11): 818–824.
[93] Arnida J-AMM, Ray A, Peterson CM, Ghandehari H. Geometry and surface characteristics of gold nanoparticles influence their biodistribution and uptake by macrophages. Eur J Pharm Biopharm 2011;77(3):417–423.
[94] Cho EC, Au L, Zhang Q, Xia Y, Xio Y. The effects of size, shape, and surface functional group of gold nanostructures on their adsorption and internalization by cells. Small 2010;6(4): 517–522.
[95] Santi M, Maccari G, Mereghetti P, et al. Rational design of a transferrin-binding peptide sequence tailored to targeted nanoparticle internalization. Bioconjug Chem 2017;28(2): 471–480.
[96] Ranalli A, Santi M, Capriotti L, et al. Peptide-based stealth nanoparticles for targeted and pH-triggered delivery. Bioconjug Chem 2017;28(2):627–635.
[97] Shukla R, Bansal V, Chaudhary M, Basu A, Bhonde RR, Sastry M. Biocompatibility of gold nanoparticles and their endocytotic fate inside the cellular compartment: a microscopic overview. Langmuir 2005;21(23):10644–10654.
[98] Cho E, Xie J, Wurm P, Xia Y. Understanding the role of surface charges in cellular adsorption versus internalization by selectively removing gold nanoparticles on the cell surface with a I2/KI. Nano Lett 2009;9(3):1080–1084.
[99] Khan J, Pillai B, Das T, Singh Y, Maiti S. Molecular effects of uptake of gold nanoparticles in HeLa cells. Chembiochem 2007;8(11):1237–1240.

[100] Villanueva A, Canete M, Roca AG, et al. The influence of surface functionalization on the enhanced internalization of magnetic nanoparticles in cancer cells. Nanotechnology 2009;20 (11):115103–115112.
[101] Murphy CJ, Gole AM, Stone JW, et al. Gold nanoparticles in biology: beyond toxicity to cellular imaging. Acc Chem Res 2008;41(12):1721–1730.
[102] Henriksen-Lacey M, Carregal-Romero S, Liz-Marzán LM. Current challenges toward in vitro cellular validation of inorganic nanoparticles. Bioconjug Chem 2017;28(1):212–221.
[103] LaBarbera DV, Reid BG, Yoo BH. The multicellular tumor spheroid model for high-throughput cancer drug discovery. Expert Opin Drug Discov 2012;7(9):819–830.
[104] Mapanao AK, Santi M, Faraci P, Cappello V, Cassano D, Voliani V. Endogenously triggerable ultrasmall-in-nano architectures: targeting assessment on 3D pancreatic carcinoma spheroids. ACS Omega 2018;3(9):11796–11801.
[105] Bourquin J, Milosevic A, Hauser D, et al. Biodistribution, clearance, and long-term fate of clinically relevant nanomaterials. Adv Mater 2018;30(19):1704307.
[106] Chan WCW. Nanomedicine 2.0. Acc Chem Res 2017;50(3):627–632.
[107] Alexander J. Confusing debriefing and defusing postnatally: The need for clarity of terms, purpose and value. Midwifery 1998;14(2):122–124.
[108] Schleh C, Semmler-Behnke M, Limpka J, et al. Size and surface charge of gold nanoparticles determine absorption across intestinal barriers and accumulation in secondary target organs after oral administration. Nanotoxicology 2012;6(1):36–46.
[109] Schmid G, Kreyling WG, Simon U. Toxic effects and biodistribution of ultrasmall gold nanoparticles. Arch Toxicol 2017;91(9):3011–3037.
[110] Lasagna-Reeves C, Gonzalez-Romero D, Barria M a, et al. Bioaccumulation and toxicity of gold nanoparticles after repeated administration in mice. Biochem Biophys Res Commun 2010;393(4):649–655.
[111] Niidome T, Yamagata M, Okamoto Y, et al. PEG-modified gold nanorods with a stealth character for in vivo applications. J Control Release 2006;114(3):343–347.
[112] Bisht S, Maitra A. Dextran-doxorubicin/chitosan nanoparticles for solid tumor therapy. Wiley Interdiscip Rev Nanomed Nanobiotechnol 2009;1(4):415–425.
[113] Danhier F. To exploit the tumor microenvironment: Since the EPR effect fails in the clinic, what is the future of nanomedicine? J Control Release 2016;244:108–121.
[114] Bertrand N, Wu J, Xu X, Kamaly N, Farokhzad OC. Cancer nanotechnology: The impact of passive and active targeting in the era of modern cancer biology. Adv Drug Deliv Rev 2014;66:2–25.
[115] Wicki A, Witzigmann D, Balasubramanian V, Huwyler J. Nanomedicine in cancer therapy: challenges, opportunities, and clinical applications. J Control Release 2014;200:138–157.
[116] Rosenblum D, Joshi N, Tao W, Karp JM, Peer D. Progress and challenges towards targeted delivery of cancer therapeutics. Nat Commun 2018;9(1).
[117] Eaton MAW, Levy L, Fontaine OMA. Delivering nanomedicines to patients: A practical guide. Nanomedicine Nanotechnology, Biol Med 2015;11(4):983–992.

3 Gold applied to nanomedicine

During the last decades, gold nanoparticles (AuNPs) have been proposed for a profusion of potential applications in medicine (Figure 3.1) due to their appealing features [1–3]. In this chapter, far from a comprehensive discussion that readers may find elsewhere, the most recent advances in the field of nanomedicine are reported [4].

Readers may notice that none of the following AuNPs is commonly employed in clinical setting [5]. Indeed, AuNPs are still not approved as therapeutics for human use because, despite the interesting preclinical results, they persist in organisms after the medical action [2, 6]. Indeed, body excretion of exogenous materials above 10 nm occurs through the hepatobiliary route, which in the case of nonbiodegradable noble metals is slow and inefficient [1, 6]. As also communicated by the Scientific Committee on Emerging and Newly Identified Health Risks (SCENIHR), exogenous materials for healthcare have to be completely cleared from the body in a reasonable amount of time, avoiding persistence in the organism that would increase the likelihood of toxicity [6–8]. This pivotal subject is discussed along the chapter. On the other hand, it is worth to notice that some metal-based nanodiagnostic are on the market because intended only for external use [4]. In particular, one is a colorimetric assay based on AuNPs conjugated to monoclonal antichorionic gonadotropin antibodies for the detection of pregnancy state by urine analysis [9]. Another one is based on AuNPs and lateral flow immunochromatographic assay for HIV detection. The last one is a tool called *Verigene*, for the identification and quantification of nucleic acid sequences in biological fluids.

This chapter comprises two main sections: Section 3.1 reports on the diagnostic use of AuNPs, while Section 3.2 describes the potential therapeutics and theranostics applications.

3.1 Diagnostics and imaging

3.1.1 Colorimetric essays

One of the most important characteristics of the noble metal nanomaterials is the LSPR (Chapter 2), which is localized at 520 nm for small gold nanospheres (AuNSs). The dependence of the plasmon band position on the degree of aggregation of the nanoparticles and from the refractive index of the surrounding medium (Chapter 2) is the basis of chemical sensing with AuNPs (Figure 3.2) [10].

Usually, aggregation of AuNSs results in a color shift of the solutions from red to violet or blue, corresponding to a characteristic red shift in the LSPR of the particles [10]. The process of aggregation is induced by the presence of chemical or biological targets that results in a change in the local refractive index, i.e. a spectral change of

https://doi.org/10.1515/9781501511455-003

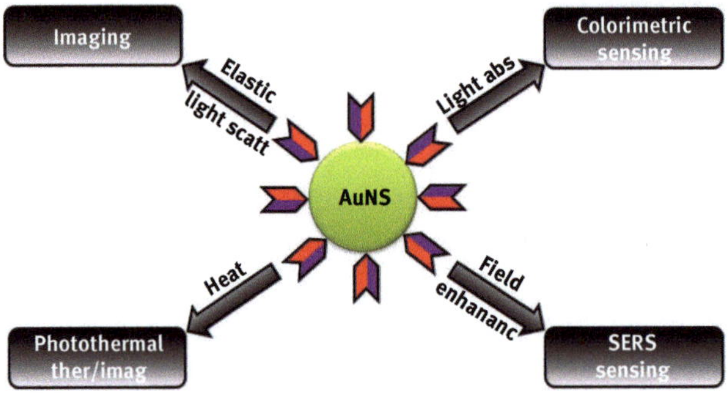

Figure 3.1: Cartoon depicting some of the most promising applications of AuNPs in medicine.

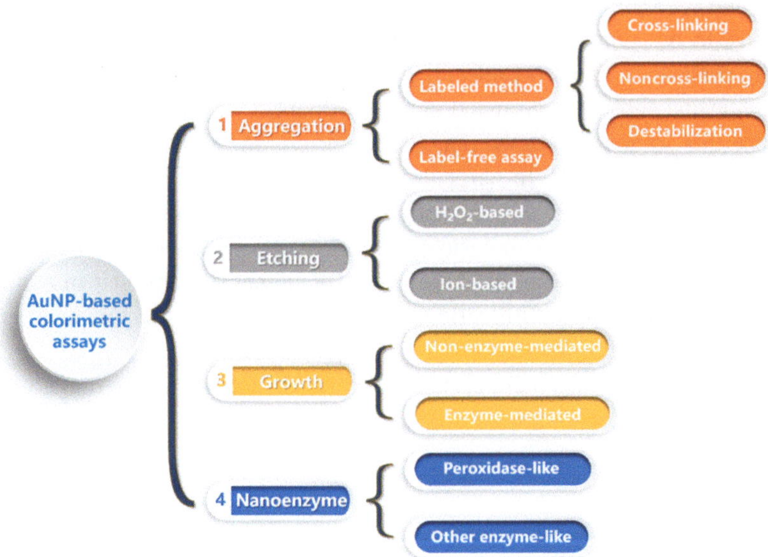

Figure 3.2: Summary of different types of gold nanomaterial-based colorimetric sensors. Reproduced with permission from C.C. Chang et al., Nanomaterials, 2019, MDPI [11].

the plasmon band position [10, 12]. Moreover, as discussed in Chapter 2, the extinction coefficient of AuNPs is very high. Indeed, the overarching goal of AuNPs-based colorimetric assays is the ability to appreciate the LSPR variation by human eye at very low concentrations of analytes [3, 13].

Gold-based colorimetric assays have found some employment in medical diagnosis (Figure 3.3), in particular for the detection of deoxyribonucleic acid (DNA), ribonucleic acid (RNA), proteins, antibodies, glucose, enzymes, and metal ions, in both aqueous and physiological media [9, 10, 12].

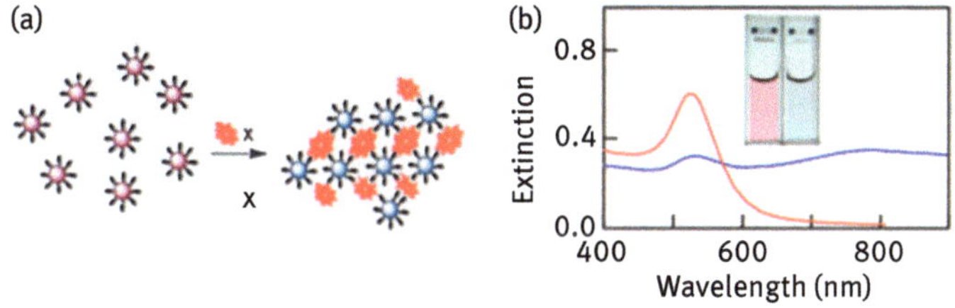

Figure 3.3: Sol-particles immunoassay. (a) Scheme for the aggregation of conjugates as a result of binding by target molecules and (b) the corresponding changes in the spectra and in sol color. Reproduced with permission from L. Dykman and N. Khlebtsov, Chemical Society Reviews, 2012, 41, 2256. ©2012, The Royal Society of Chemistry [14].

In order to induce the color change of AuNPs solutions, i.e. the coupling of LSPRs, the surface of the nanostructures is often modified with recognizing molecule [11]. One example is by using nucleotide sequences [15]. In this case, the aggregation of oligonucleotide-modified AuNPs is due to the presence of a complementary target oligonucleotide in solution [14, 15]. The AuNPs aggregation results in a shift of the solution color from red (LPSR maximum for single AuNSs in solution is at 520 nm, see Chapter 2) to blue [16]. It is worth to notice that the sensibility of this approach can reach the picomolar, and may be further improved if AuNPs are covered by a silver shell [17, 18].

Sensitivity of the diagnostic techniques is a pivotal subject of increasingly interest, in particular for the early detection of some disorders. For example, some neoplasms produce and release peculiar antigens at the early stage of development [17]. In this regard, gold-based devices modified with appropriate agents have successfully demonstrated their potentiality [17].

It is interesting to notice that spherical AuNPs are generally employed in colorimetric assays, because of the readiness of data interpretation. Investigation on the controlled aggregation of nanostructures with various geometries, among which rods or triangles, are in progress, because they may hold more information due to their anisotropic geometry [10].

3.1.2 Surface-enhanced Raman scattering

There are a number of surface-enhanced spectroscopies that result from LSPR stimulation. In this section the focus is on surface-enhanced Raman scattering (SERS) [19, 20]. For a complete overview of the subject, see also some other interesting reviews [19, 20]. Raman spectroscopy is a powerful tool for the identification of many substances, due to the very specific spectral vibrations of samples [21–23]. In Raman spectroscopy, photons are not absorbed by the analyte but inelastically scattered. The fundamental selection rule in Raman spectroscopy is the polarizability modification of the molecule during the vibration modes and the intensity of Raman signals depends on the fourth power of the local electric field, that is, the signals are often too low to be collectable. It is interesting to remember that the fundamental selection rule for infrared (IR) spectroscopy is the modification of the dipole moment of molecules during the intramolecular vibration.

SERS spectroscopy is based on the enhancement of Raman scattering from molecules in proximity of (adsorbed on or linked to) metal surfaces [24]. The enhancement can reach values of up to 10^{14} if appropriate nanoparticles are employed, and decreases very rapidly when the distance between the molecule and the metal surface increases [25]. SERS resulting by the conjugation of AuNPs and molecular barcodes has been proposed for medical imaging and anticounterfeit detection [26, 27]. There are two primary mechanisms of Raman enhancement (for a complete discussion see Section 2.1.6): a chemical enhancement and an electromagnetic (EM) one [20]. About the latter, the metal acts as a charge-transfer mediator to the analyte when the energies of the highest occupied molecular orbital (HOMO) and the lowest unoccupied molecular orbital of the adsorbate are symmetrically disposed with respect to the Fermi level of the metal. Thus, the half energy of the excitation light is employed in the electron transition from the Fermi level to the HOMO level. On the other hand, the LSPR produces a secondary electric field, which enhances both the incident EM field and the scattered Raman emission. Irregularities on AuNPs surface usually increase the Raman enhancement improving the sensibility of the technique to the single molecule [28]. It is important to note that (i) the enhancement of the field is particularly strong in the presence of "hot spots," and (ii) in the range of 17–80 nm the Raman spectral intensities are independent from the size of AuNPs [29]. The hot spot is a peculiar volume of space around one (or more) nanoparticle, in which the Raman signals are significantly increased. Indeed, very intense SERS signals can be collected from aggregated particles or geometries with plenty of edges, such as rods, triangles, and stars [28]. It is worth to remember that SERS is a more sensible technique with respect to fluorescence detection because the background noise is minimized and the sharp signals easily collectable [30].

AuNPs have been evaluated for the detection of many targets, including metal ions, small organic compounds, proteins, DNA, RNA, and cells [28]. SERS-based assays are usually tested in vitro, including in lab-on-chip set-up, or in living cells [28]. An example of the first approach relies on the microarray-based Raman spectroscopy for the detection of protein binding and kinase activity [31]. Another example is reported in Figure 3.4, in which an antiphosphoserine-biotin is employed to recognize the phosphorylation on a surface. Avidin-fluorescein labels the immobilized peptides (fluorescein acts as a Raman-active probe), and peptide-stabilized AuNPs interact with the system by a standard avidin–biotin linkage.

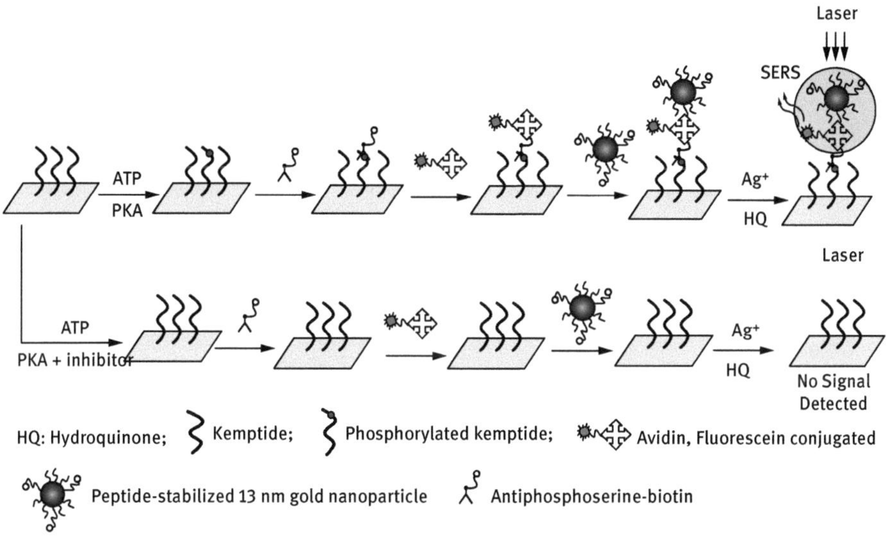

Figure 3.4: Schematic representation of SERS analysis of phosphorylation on surfaces. The phosphorylation event is recognized by SERS measurements thank to the enhancement of the AuNP covered by a silver layer. ATP, adenosine-5′-triphosphate; HQ, hydroquinone; PKA, protein kinase. Reproduced with permission from D.T. Nguyen, D.J. Kim and K.S. Kim, Micron, 2011, 42, 207. ©2011, Elsevier [32].

In order to enhance the SERS signal, silver is deposited on AuNPs linked to the surface by reduction of Ag^+. By this approach, up to 10 fg of standard proteins can be recognized. Another example is a tool for circulating tumor cells detection comprising AuNPs conjugated with specific peptides able to detect the targets in mouse and human blood samples (Figure 3.5) [33].

Wang and coworkers demonstrated its good specificity in recognizing the targets in blood with sensitivity of 5–50 cells/ml of blood [33]. Both modified and unmodified AuNPs have also been proposed for the SERS detection of analytes inside living cells [19, 34]. Moreover, Huang and coworkers have demonstrated that some cancer cells can be discriminated by their membrane decoration with antibody-modified gold nanorods (AuNRs) [34]. On one hand, Kneipp and coworkers discriminated malignant from nonmalignant cells by employing unmodified AuNPs [19]. Interestingly, compounds and components of cells can differently interact with gold nanoprobes depending on their malignancy, and this results in modification of the Raman spectra [35]. For example, in Figure 3.6 the peak at 731 cm^{-1}, which is assigned to adenine, is stronger in cancer cells. In addition, a peak at 398 cm^{-1} is observed only in cancer cells. These spectral differences can be useful in cancer diagnosis.

It is worth to notice that AuNPs-driven SERS and electron microscopy techniques can be correlated in order to investigate time-dependent chemical changes in cells and bacteria [36, 37]. These investigations provide in-depth knowledge of cellular processes

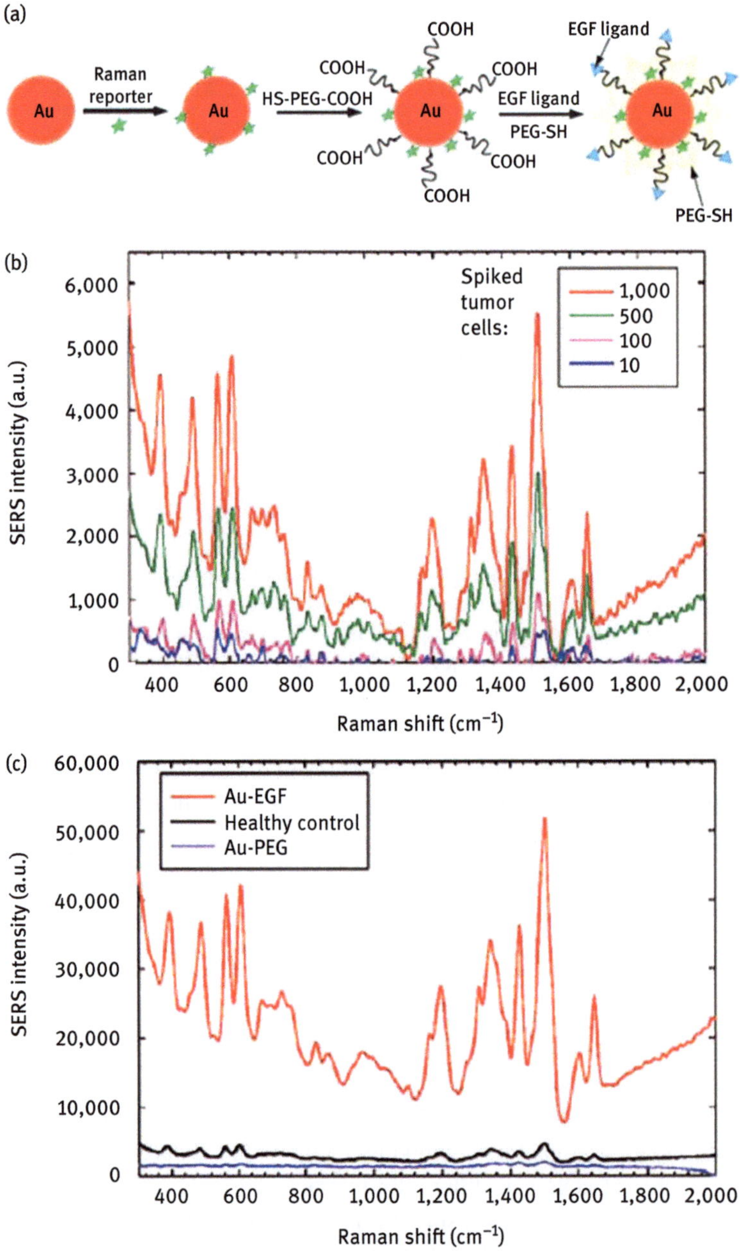

Figure 3.5: SERS detection of circulating tumor cells in patient blood samples. (a) Schematic illustration of SERS NP and their conjugation with epidermal growth factor (EGF) peptides. (b) SERS spectra of different numbers of Tu212 cancer cells spiked into mouse white blood cells. (c) SERS spectra of blood sample from a patient incubated with targeted and nontargeted SERS NP, as well as a blood sample from a healthy donor incubated with targeted SERS NP. The SERS NP can detect

at the molecular level, among which chemical composition gradients in cells and intracellular location of nanostructures.

In summary, SERS is a promising technique for imaging (Figure 3.7) and investigation at molecular level of cells processes. On the other hand, signals can be often complicated by the heterogeneity in the surrounding cellular matrix or impurities surrounding the particles, whose Raman signals can also be enhanced.

3.1.3 Optical and optoacoustic imaging

AuNPs present strong radiative properties (light absorption, scattering, and emission) compared to bulk materials or organic molecules [38]. These features, taken together with AuNPs chemical and physiological behaviors, enable gold nanomaterials especially promising in biomedicine [4, 39]. In this section, some interesting optical and optoacoustic imaging methods that involve AuNPs are discussed, among which are light scattering, two-photon fluorescence, photoacoustic, and fluorescence imaging. Other imaging techniques, including X-ray computed tomography and positron emission tomography, are comprehensively discussed elsewhere [4, 40–42].

AuNPs strongly scatter light if irradiated on their LSPR (Chapter 2) because they exhibit big scattering sections that are many times higher respect the ones regarding dyes [43]. Indeed, AuNPs larger than 10 nm in diameter can be readily visualized even at the single particle by dark field (DF) microscopy with a simple optical microscope [44, 45]. In addition, the light scattering of nanoparticles is photostable without blinking, enabling long-time investigations. DF images can be collected by both functionalized and nonfunctionalized AuNPs on surfaces or inside cells (Figure 3.8) [46].

For example, cervical cancer cells have been imaged and recognized by DF microscopy after labeling with 12 nm anti-EGFR-modified AuNPs due to the specific binding of the particles to the EGFR proteins on the surfaces of the cells [47]. One of the interesting advantage of DF microscopy is related to the high-contrast with true color, resulting in a potential multiplex detection approach by using AuNPs with different size and shapes [48]. Indeed, it is useful to remember here that the LSPR position in the spectrum is associated with the geometry of the particles (Chapter 2). Currently, DF microscopy based on the scattering features of receptor-functionalized AuNPs, in particular rods and shells, resulted quite promising for cancer imaging and detection

---

Figure 3.5 (continued)
circulating tumor cells with a sensitivity of 5–50 cells per ml blood. The strong signals from a cancer patient indicates highly specific and sensitive detection of circulating tumor cells in the blood system. a.u., arbitrary units; HS-PEG-COOH, alpha-thiol-omega-carboxy poly(ethylene glycol); PEG-SH, alpha-thiol poly(ethylene glycol); Tu212, hypopharyngeal carcinoma. Reproduced with permission from E.C. Dreaden, A.M. Alkilany, X. Huang, C.J. Murphy and M.A. El-Sayed, Chemical Society Reviews, 2012, 41, 2740. ©2012, The Royal Society of Chemistry [30].

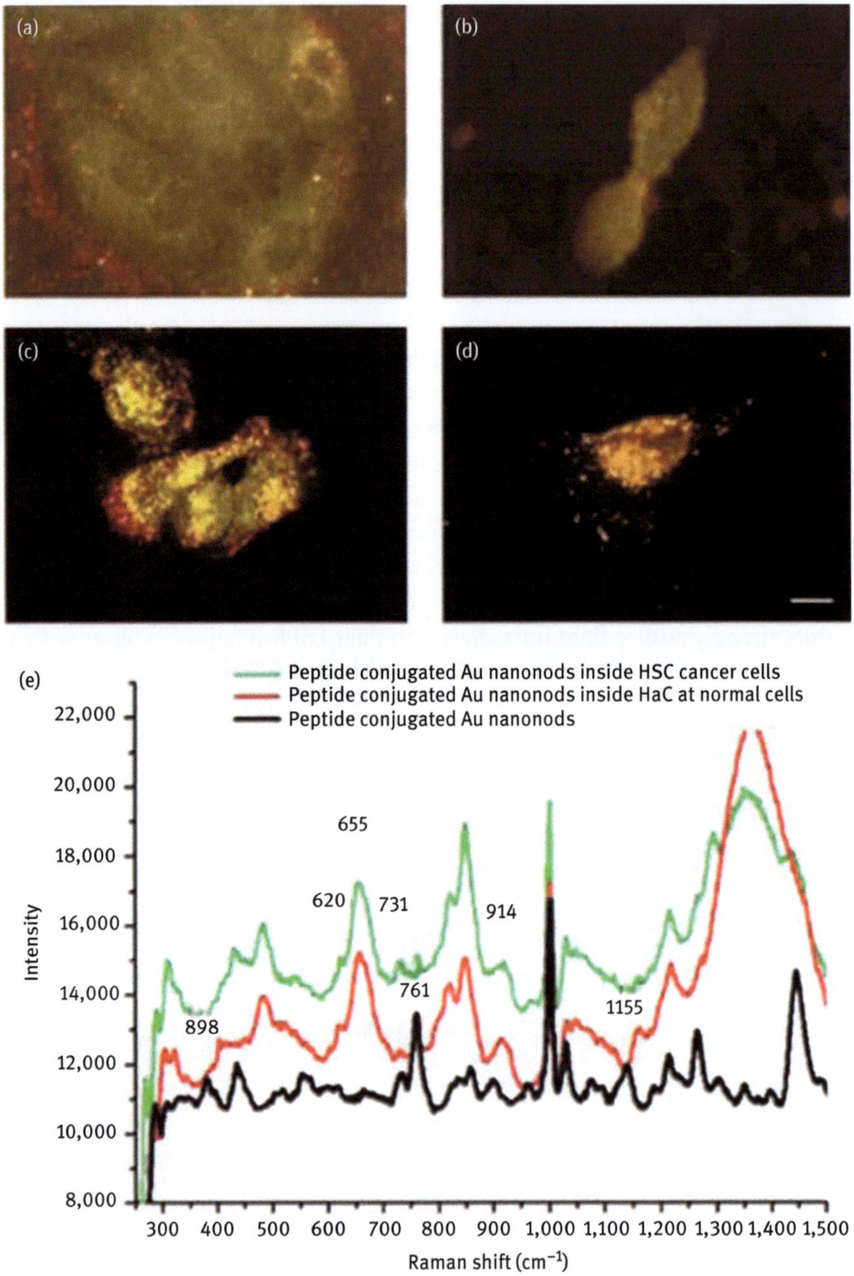

Figure 3.6: Dark field (DF) light scattering images of cells (a and c) and cancer cells (b and d) after incubation for 2 h with AuNRs and peptide-conjugated AuNRs, respectively. (e) Both type of cells shows some differences in their Raman signals. Reproduced with permission from D.T. Nguyen, D.J. Kim and K.S. Kim, Micron, 2011, 42, 207. ©2011, Elsevier [32].

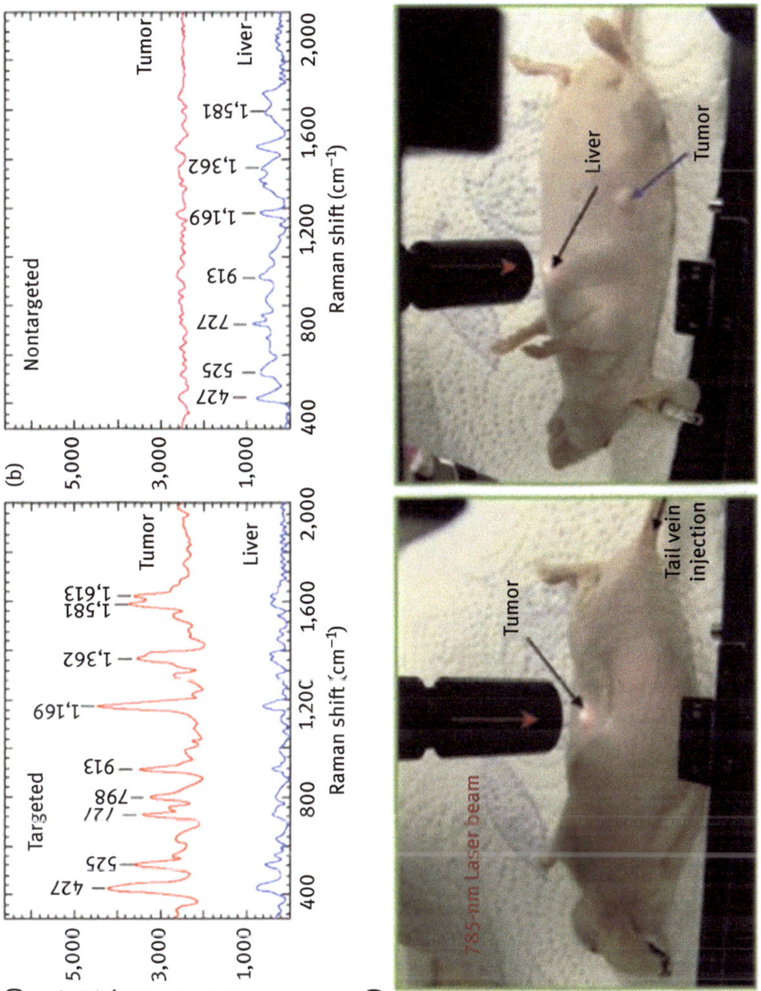

Figure 3.7: In vivo cancer targeting and SERS detection by using single-chain variable fragment antibody (ScFv) antibody conjugated gold nanoparticles that recognize the tumor biomarker epidermal growth factor receptor (EGFR). (a, b) SERS spectra obtained from the tumor and the liver locations by using targeted (a) and nontargeted (b) nanostructures. Nude mice bearing human head-and-neck cell carcinoma (Tu686, 3 mm diameter) received 90 ml of ScFv EGFR-conjugated SERS tags or PEGylated SERS tags (460 pM). The particles were administered via tail vein single injection. SERS spectra were taken 5 h after injection. (c) Photographs showing a laser beam focusing on the tumor site or on the anatomical location of the liver. In vivo SERS spectra were obtained from the tumor site (top spectra) and the liver site (bottom spectra) with a two second signal integration and at 785 nm excitation. The spectra were background subtracted and shifted for better visualization. Reproduced with permission from X. Qian, X.H. Peng, D.O. Ansari, Q. Yin-Goen, G.Z. Chen, D.M. Shin, L. Yang, A.N. Young, M.D. Wang and S. Nie, Nature Biotechnology, 2008, 26, 83. ©2008, Nature [35].

(Figure 3.8). Moreover, the scattering features of AuNPs can be also employed for real-time observation of intracellular processes at single-particle event [49].

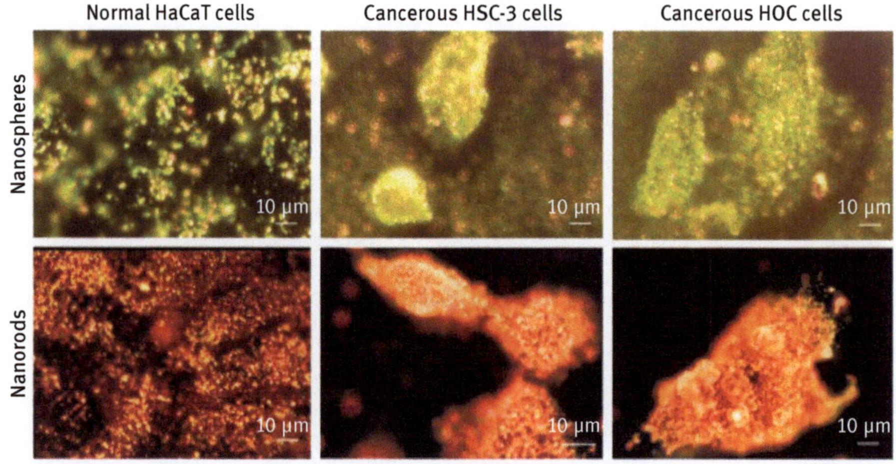

Figure 3.8: Cancer diagnostics using gold nanorod-enhanced light scattering. Optical DF microscopy of human keratinocytes human keratinocytes (HaCaT) cells and cancerous HSC and HOC cells incubated with anti-EGFR antibody conjugated AuNSs (top panels, left to right). Optical DF microscopy of normal HaCaT cells and human oral squamous cell carcinoma (HSC and HOC) cells incubated with anti-EGFR antibody-conjugated AuNRs (lower panel, left to right). Anti-EGFR-conjugated AuNP specifically bound to cancer cells, scattering strongly under DF microscopy and thus enabling detection of malignant cells. Reproduced with permission from E.C. Dreaden, A.M. Alkilany, X. Huang, C.J. Murphy and M.A. El-Sayed, Chemical Society Reviews, 2012, 41, 2740. ©2012, The Royal Society of Chemistry [30].

Some AuNPs geometries, such as rods or cubes, resulted interesting for two-photon luminescence (TPL) technique if coupled with pulsed (fs) near-infrared (NIR) laser light [38]. In general, TPL was investigated for imaging of cancer cells both in vitro and in vivo. TPL technique has the advantage of high spatial resolution and reduced background noise when compared to confocal microscopy [50, 51]. On the other hand, pulsed NIR laser irradiation employed for TPL can also induce local photothermal (PT) effects that affect the cells viability [52, 53]. The cellular localization of AuNPs can be easily differentiated due to the high contrast associated with TPL imaging [51]. Indeed, the intensity of photoluminescence signals collected by gold nanocubes is 600 times stronger (100 times for nanorods) than the emission of single fluorescein isothiocyanate molecule. Due to the high emission rates of AuNPs, TPL technique was investigated for in vivo: (i) detection of circulating tumor cells [54], and (ii) biokinetics evaluation of nanomaterials in animal models [54]. An example of the first is reported by He and colleagues: circulating tumor cells (CTC)-mimetic leukemia cells were injected in the blood stream of live mice, followed by injection of folate-modified gold nanorods. Then, the authors were able to detect single cancer cells in the vasculature of the mouse ear by TPL [55].

Photoacoustic imaging (PAI) is based on the use of laser-induced heating of materials, and relies on the detection of acoustic waves generated by the thermal expansion of the media that surrounds AuNPs [56]. PA imaging is a promising diagnostic technique that combines the high contrast of optical imaging with the deep tissue penetration of ultrasound imaging [56]. In particular, this technique is very interesting when coupled with NIR-absorbing AuNPs, because of the good penetration of light in tissues in the biological window [57–59]. Widespread applications of PAI have been investigated in recent years, among which tumor imaging (Figure 3.9) [56].

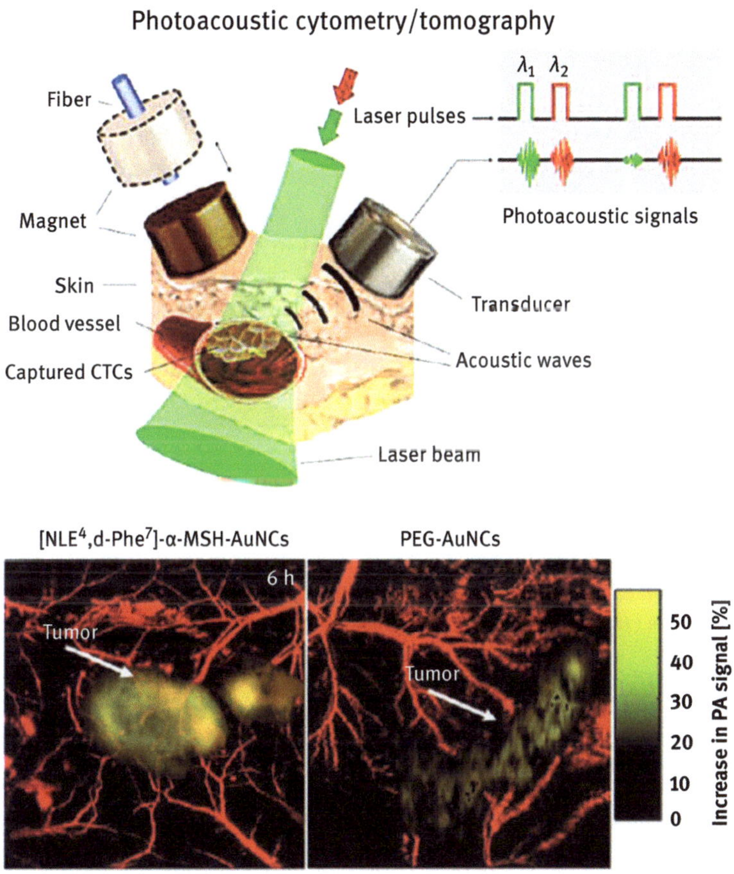

Figure 3.9: Photoacoustic cytometry/tomography: pulsed laser excitation of cells/tissues labelled with AuNPs can be used to detect or sequester circulating tumor cells (CTC) or to noninvasively image/diagnose/stage tumors and guide surgical procedures. [NLE^4, d-Phe^7]-α-MSH-AuNCs, 4-norleucine, 7-D-phenylalanine-α-melanocyte-stimulating hormone-gold nanocages. Reproduced with permission from E.C. Dreaden and M.A. El-Sayed, Accounts of Chemical Research, 2012, doi: 10.1021/ar2003122. ©2012, American Chemical Society [38].

For example, Emelianov and coworkers demonstrated the possibility to employ PAI to detect subcutaneous tumor in mice with antibody-conjugated AuNSs [60]. Other examples regard the use of gold nanocages (Figure 3.10) to image blood vessels of mouse brain or gold nanostars for the imaging of the rat lymphatic system [30].

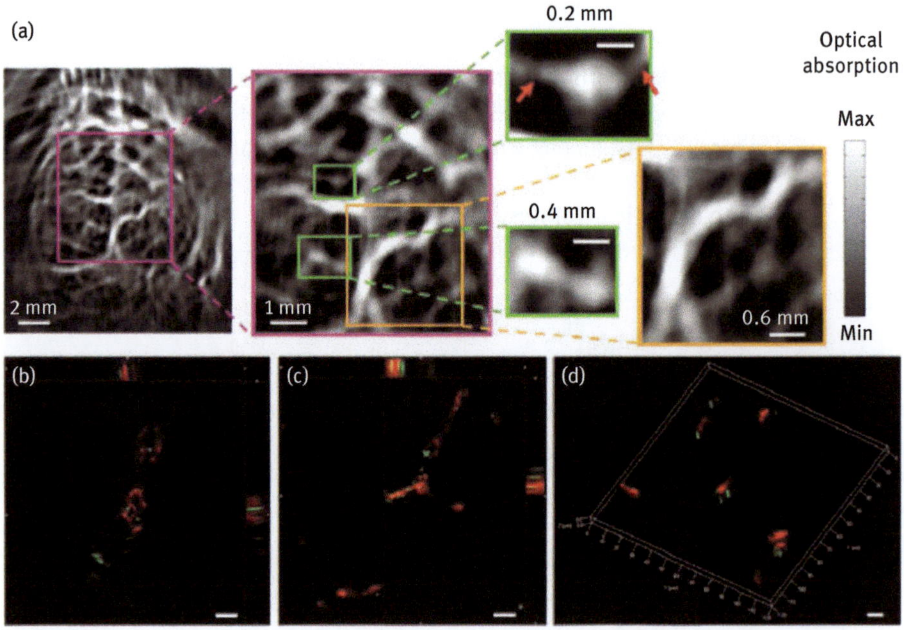

Figure 3.10: PA imaging of blood vessels in the mouse brain using gold nanocages. (a) PA image of large (yellow-framed picture) and small (green-framed picture) blood vessels of a mouse brain 2 h after intravenous injection of PEG-conjugated nanoparticles. (c–d) Optical images of mouse brain vessels 2 h after injection with PEG-conjugated nanoparticles. Blood vessels are stained with anti-CD31 antibody (red). Nanoparticles are imaged by DF scattering microscopy (green). Nanoparticles enhanced the PA signals of blood vessels in the mouse brain, revealing a clear and detailed structure vasculature as small as 100 mm in diameter. Reproduced with permission from E.C. Dreaden, A.M. Alkilany, X. Huang, C.J. Murphy and M.A. El-Sayed, Chemical Society Reviews, 2012, 41, 2740. ©2012, The Royal Society of Chemistry [30].

It is important to remember that AuNPs proposed for PAI are usually rod-, cage-, or star-shaped structures that suffer from metal persistence in organism after the potential action [2, 6]. On one hand, a novel approach for the composition of PAI probes rely on the conjugation of dyes with plasmonic ultrasmall AuNPs. These materials have demonstrated an impressive versatility together with a good signal-to-noise ratio and (bio)degradation/excretion of the building blocks [7, 58, 59, 61].

If AuNPs have a size below 2 nm (cluster-size regime), the plasmonic features are usually lost and other behaviors appear, including fluorescence and catalytic effects (Figure 3.11).

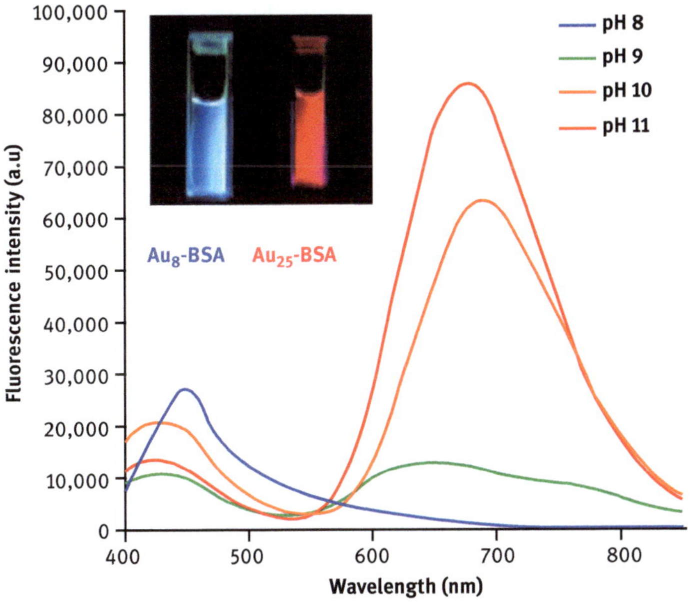

Figure 3.11: Emission fluorescence spectra (λ_{exc} = 360 nm) of AuNSs-BSA samples prepared at different pH. Reproduced with permission from X. Le Guével, B. Hötzer, G. Jung, K. Hollemeyer, V. Trouillet and M. Schneider, The Journal of Physical Chemistry C, 2011, 115, 10955. ©2011, American Chemical Society [62].

The fluorescence is likely to be due to the transition of moleculelike electronic levels when subnanometer sizes are smaller than the Fermi wavelength [63]. In particular, the emission may be related to the gold interband transition $5d^{10}$ to 6sp and/or ligand–metal charge transfer transition [63]. Interestingly, gold clusters do not suffer from the typical drawbacks of fluorophores or quantum dots, among which are photostability and toxicity/blinking. Recent advances in the synthesis of metal clusters have led to the optimization of highly fluorescent and water-soluble platforms intended for biomedical application [63]. For example, an immunoassay for the detection of human IgG based on an antibody-modified gold clusters was developed and evaluated [64].

3.2 Therapeutics

Gold nanomaterials have attracted increasing interest for the potentiality to revolutionize the treatment of many diseases, especially neoplasms. On the other hand, a huge imbalance between proposed nanoplatforms and approved ones for clinical applications is present. This disequilibrium strongly affects AuNPs, which present no-approved platform and very few candidates in clinical trials because of the issue of metal persistence [6]. Thus, readers have to consider that the topics and platforms here reported are still far from the clinical settings [2, 6]. In this section, a focus is placed on photothermal agents and delivery nanosystems because of their good translational potential. Readers may refer elsewhere if interested in other potential applications of AuNPs [4, 39].

3.2.1 Photothermal therapy

HyperThermia (HT) is an ancient therapeutic procedure consisting in raising the temperature of diseased body areas or of whole body. HT is commonly defined as a temperature increase up to 45 °C, as higher temperature treatments fall into the thermal ablation techniques [65]. For the treatment of cancer-related diseases, HT is either supplied alone or in support to other therapies, enhancing the therapeutic action. Interestingly, HT may not significantly affect cultured cells viability, while induce a significant effect in vivo due to the vasculature of solid tumors, the highly acidic environments of neoplasms, and the usual marked hypoxia that cancer cells suffer [66]. The effectiveness of HT treatment against tumors has been ascribed to several temperature-dependent biological mechanisms, including impaired DNA double strand break repair, misregulation of gene expression, increased mitochondrial reactive oxygen species (ROS) production, cytoskeleton disruption resulting in altered intracellular transport, antitumor immune response activation, and many others [67–72]. HT treatment of solid tumors can be locally supplied by irradiation (photothermal therapy, PTT) and can be assisted by plasmonic nanoparticles (Figure 3.12).

Commonly, the nanostructures employed for PT investigations are AuNPs that exhibit LSPRs at NIR frequencies, i.e. anisotropic structures among which nanostars and nanorods [73–76]. It is worth to remember that the employment of NIR-light in PTT is pivotal as these frequencies allow for high-depth penetration of the light in tissues (first and second biological windows) [8, 57]. Besides their production by the employment of the quite cytotoxic hexadecyltrimethylammonium bromide (CTAB), gold nanorods are easy to prepare (Chapter 1) and present tunable LSPRs in the visible-NIR (Chapter 2). These behaviors have supported their evaluation on cancer cells and animal models with some extent of success [77]. Remarkably, gold nanorods can be exploit to jointly combine PTT with PAI, resulting in interesting theranostics platforms [77].

Another class of NIR-absorbing AuNPs that have been successfully evaluated as PT agents are the gold nanocages [78]. Gold nanocages are cubic-shape hollow structures (Chapter 1) with tunable plasmonic optical properties in the range of 400–1,200 nm, depending on their wall thickness and edge length [79]. For example, Xia and coworkers used targeted gold nanocages with plasmonic absorption maxima at 810 nm to ablate SK-BR-3 breast cancer cells (Figure 3.13) [80].

Noticeably, gold nanoshells have also emerged as an excellent photothermal transducer, as their NIR optical response can be easily tuned by adjusting the diameter-to-

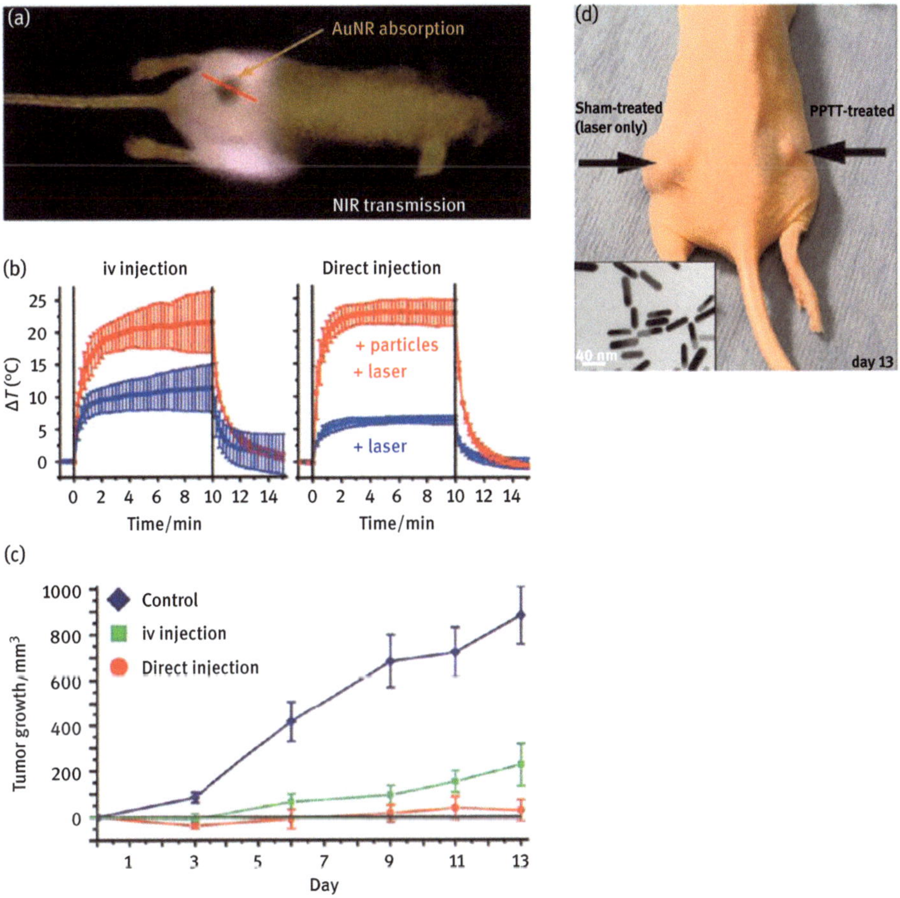

Figure 3.12: (a) NIR transmission image of tumor-bearing mice (rear flank) showing gold nanorods specifically accumulated at the tumor site (dark, highly absorbing region). (b) Thermal transient measurements from the tumor center during NIR laser exposure of tumors treated by intravenous (left) and intratumor (right) administration. High thermal contrast is observed between particle treated mice (red curves) and controls treated by laser only (blue). (c) Tumor regression following intravenous (green) and intratumor (red) laser PTT when compared to untreated controls (blue). No detectable disease was observed at day 13 in >50% of the interstitially treated group and 25% of the intravenously treated group. (d) Tumor growth resorption in a weakly responding mouse treated by intratumor nanorods injection (electron microscopy, inset) and laser PTT versus control (laser alone) treatment. Reproduced with permission from E.C. Dreaden and M.A. El-Sayed, Accounts of Chemical Research, 2012, doi: 10.1021/ar2003122. ©2012, American Chemical Society [38].

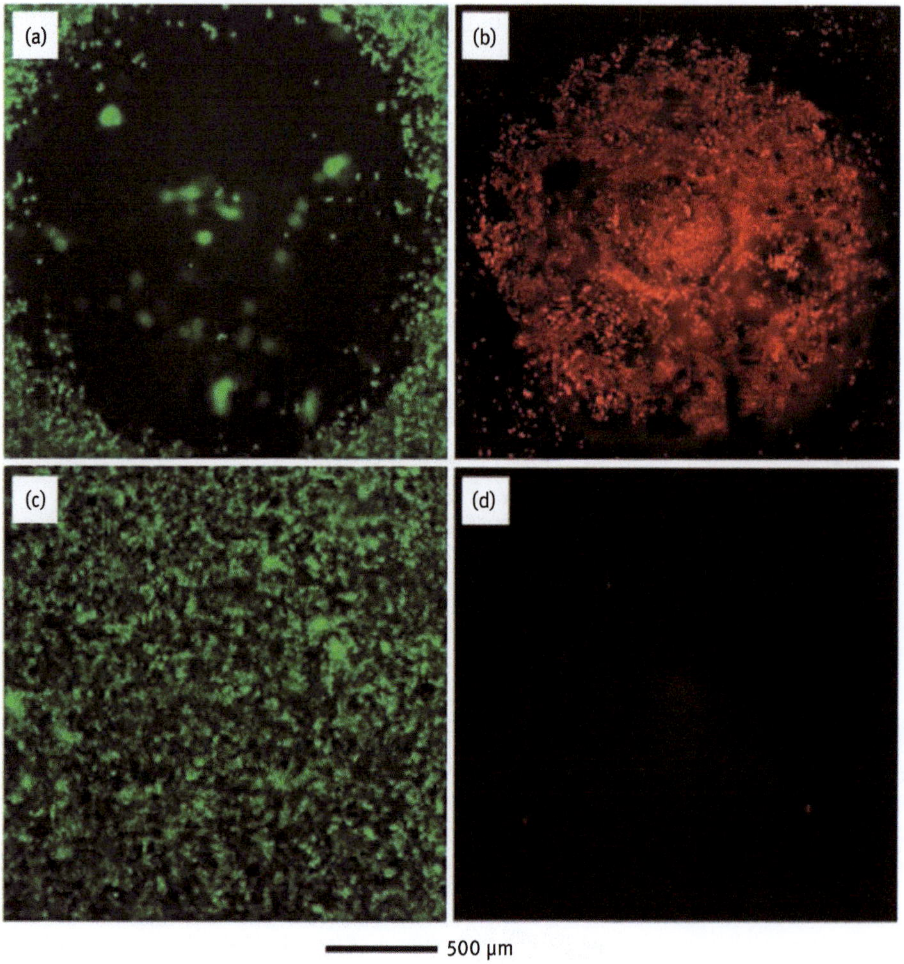

Figure 3.13: SK-BR-3 breast cancer cells that were treated with modified AuNSs and then irradiated by an 810 nm laser at a power density of 1.5 W/cm^2 for 5 min. A well-defined circular zone of dead cells was revealed by (a) calcein AM assay, where green fluorescence indicates live cells, and (b) ethidium homodimer-1 (EthD-1) assay, where red fluorescence indicates dead cells. In a control experiment, cells irradiated under the same conditions but without incubating with modified AuNSs maintained viability, as indicated by (c) calcein fluorescence assay, and (d) the lack of intracellular EthD-1 uptake. Reproduced with permission from S. E. Skrabalak, J. Chen, L. Au, X. Lu, X. Li, Y. Xia, Advanced Materials, 2007, 19, 3177. ©2007, Wiley-VCH [81].

shell thickness ratio [82]. For example, nanoshells have been evaluated in combination with magnetic systems in order to ablate various cancerous cell lines and have been successfully tested in animal models [30]. On one hand, it is interesting to notice that only nanoshells have reached the clinical trial evaluation, whereas all the other AuNPs are still confined to preclinical stage despite their excellent light-to-heat conversion performances [83].

The actual failure of clinical translation of AuNPs-mediated PTT is mainly ascribed to body persistence concerns [84]. Indeed, the common size of AuNPs showing an LSPR in the NIR is up to 150 nm [82]. On the other hand, nonbiodegradable noble metals are slowly and inefficiently excreted by the hepatobiliary route [1, 6]. Remarkably, as also communicated by SCENIHR, therapeutics have to avoid persistence in organisms and to be completely cleared from the body in a reasonable amount of time [6]. A commonly adopted approach to avoid noble metal persistence is to reduce AuNPs size below the threshold for renal clearance, i.e. ultrasmall nanoparticles (USNPs) [3]. Furthermore, the maximum light-to-heat transduction is obtained by NPs smaller than 5 nm, as the photothermal conversion efficiency is size dependent [85]. Unluckily, LSPRs of excretable gold USNPs are in the UV/visible region, severely limiting their potential application in PTT as far from the first biological windows. In this regard, some gold nanorods can partially avoid this issue. Indeed, at least one of their dimensions can be tailored in the ultrasmall range, suitable for renal excretion, while maintaining the LSPRs in the NIR [86]. However, gold nanorods reshapes into spheres during laser irradiation, preventing the possibility of multiple PTT series and resulting in nonexcretable structures [87–89]. Strategies to preserve nanorods morphology, such as silica coating, have been successfully adopted, albeit increasing their size [90]. An innovative approach for the design of nanoarchitectures (NAs) that jointly combine NIR-triggered PTT and noble metal excretion has been recently introduced: the ultrasmall-in-nano [8]. The first AuNPs developed by this approach are called *passionfruit-like NAs* and they are composed by noble metal USNPs embedded in biodegradable silica nanocapsules [23, 91] The PTT efficacy of NAs have been evaluated in significant 3D models of pancreatic carcinoma and produced very promising results [8]. The NAs biosafety and biokinetics have also been fully evaluated together with other potential single/combined applications, denoting their interesting potentiality for translational research [7, 58, 59, 61, 92, 93].

In summary, AuNPs-driven PTT is one of the most promising breakthroughs for advanced treatment of oncological patients both alone or in combination with other treatments. On the other hand, some serious challenges have to be dealt with before the translation in clinical settings, including: (i) metal persistence, (ii) standardization of the materials and protocols, and (iii) scaling up and automatization of the nanomaterials production [2, 6, 94].

3.2.2 Delivery nanosystems

The conjugation of chemotherapy with nanotechnology has already introduced in the clinical setting novel and efficient tools with increased target specificity and reduced adverse effects on the organisms [7, 95]. The enhanced action of these organic nanomaterials rely on the improved amount of drug available at the required site. This scientific advancement is due to some intriguing features of nanomaterials, among which the high drug loading ability, and the overall size in the nanometer range [6].

Beyond drug delivery, noble metals are essential for the development of innovative or combined/enhanced treatments in clinical settings that will promote another revolution in the care of several diseases [2, 91].

Generally, drugs are loaded in or on AuNPs by covalent conjugation or noncovalent interactions to both the coating or the metal surface (Figure 3.14) [96].

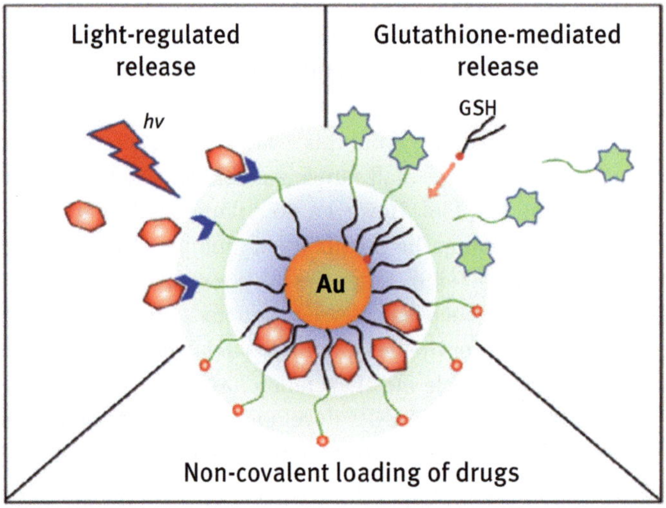

Figure 3.14: The image shows the three different types of triggers for the release of payloads from AuNP. GSH, glutathione; *hv*, light. Reproduced with permission from by C.K. Kim, P. Ghosh and V. M. Rotello, Nanoscale, 2009, 1, 61. ©2009, The Royal Society of Chemistry [97].

The noncovalent approach usually exploits active drugs and offers a good releasing efficiency, while the covalent one requires intracellular processing of a prodrug [96]. On the other hand, the release is better controlled with the covalent approach. Indeed, drugs loaded by adsorption may suffer from nonspecific release [96]. Drug release can be triggered by internal (endogenous) or external (exogenous) stimulus, such as different concentrations of molecules in different compartments of cells or tissues or by irradiation (Figure 3.14) [98, 99].

Nanocarriers able to respond to biological stimuli are of special interest as they are exclusively triggered by endogenous features, including pH, temperature, and redox microenvironment [98, 100]. For example, extracellular pH in solid neoplasms is more acidic (pH = 5.5–6.5) than in physiological conditions (about 7.4) [101]. In addition, the intracellular glutathione (GSH) level in tumor cells are 100–1000-fold higher than in normal tissues and, usually, the temperature of the organism is higher when tumor or inflammation or infection processes are in action [98, 101]. On one hand, several endogenously triggered nanosystems – mainly based on PEG-encapsulated AuNPs – have been developed and successfully evaluated in living cells.

For example, Rotello and colleagues developed an efficient tool for gene delivery in mammalian cells by employing cationic AuNPs [102]. The particles interact with DNA due to electrostatic interactions to compose a nanoparticle–DNA complex that is internalized by mammalian cells. The release mechanism relies on the unpackaging of DNA from the cationic AuNPs caused by the intracellular GSH [103]. Anionic GSH binds to the cationic gold core via disulfide exchange reactions on the complex in the intracellular environment, reducing the overall positive charge and consequently decreasing the DNA affinity for AuNPs. Another example based on the GSH endogenous triggering is shown in Figure 3.15 [100]. Here, AuNPs were covered by a monolayer of tetra(ethylene glycol)lyated cationic ligand (TTMA) and thiolated bodipy dyes (HSBDP). The TTMA ligand generates a cationic surface that

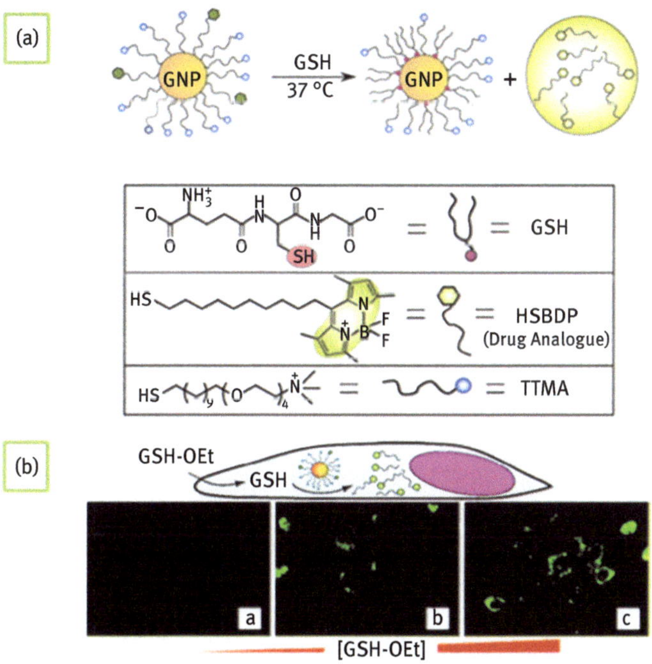

Figure 3.15: GSH-triggered drug releases in living cells. (a) Scheme of the AuNPs carrier and of the GSH-mediated surface monolayer exchange reaction, which releases the payload. (b) Schematic representation and fluorescence images when using GSH-OEt as a stimulus to release HSBDP from AuNPs. The GSH-OEt concentrations are 0, 5, and 20 mM in panels A, B, and C, respectively. In this figure GNP is for Gold Nanoparticle. Reproduced with permission from P. Ghosh, G. Han, M. De, C. K. Kim and V.M. Rotello, Advanced Drug Delivery Reviews, 2008, 60, 1307. ©2008, Elsevier [100].

enhance cellular uptake. These nanosystems are not fluorescent because the bodipy fluorescence is quenched by the proximity to the AuNPs surface. On the other hand, the increase of intracellular GSH results in a fast release of HSBDP, which promptly diffuse in the cytoplasm and recover the fluorescence yield (Figure 3.15).

It is worth to remember that the different intra/extracellular GSH concentration was also employed for the release of hidden form of drugs (prodrugs) delivered in the targets by nanomaterials [98].

Enzyme-triggering is another interesting approach for the endogenously controlled release of therapeutics. For example, Decher and coworkers developed a platform composed by a gold core covered by functionalized polyamine polymers [104]. The outer monomer building block, *N*-methacryloyl-glycyl-DL-phenylalanyl-leucyl-glycyl doxorubicin (Ma-Y-Dox), contains a nontoxic prodrug form of doxorubicin (Dox), a potent DNA intercalation agent. The spacer Y between the polymer and Dox is designed for specific cleavage by the lysosomal enzyme cathepsin. The specificity of the release mechanism relies on the exclusive release of the Dox through the enzymic digestion of the spacer, occurring in the endocytic pathway.

Endogenous systems have attracted a great deal of attention because of their peculiar ability to respond directly to a specific molecular or biological stimulus. Unfortunately, despite the steric shielding effect of the encapsulating monolayer, thiol-disulfide exchange phenomena with proteins (such as blood proteins) or unexpected release processes can occur before reaching the target. Thus, carriers may show altered bioavailability and pharmacokinetic profiles [100]. These are potential drawbacks to consider during the preclinical stage. Indeed, a full set of nanomaterials evaluations have to be produced in order to increase their success of translation to the clinical settings [105].

NIR photothermal-induced release from AuNPs is one of the most investigated approach for the exogenous-triggered release of molecules. Indeed, AuNPs features for light-to-heat conversion have been widely assessed, especially for the development of PTT agents (Section 3.2.1). Besides PTT, local heating resulting from AuNPs can be employed for the controlled release of dyes or therapeutic agents conjugated to the nanoparticles. Generally, PT release relies on (i) direct desorption of molecules from the nanoparticles, or (ii) desorption caused by conformational changes of thermally responsive envelopes, such as polymers [1]. In the following section, mechanisms, advantages, and disadvantages of these methods are discussed.

The direct desorption of agents from AuNPs usually results from weak interactions between the molecular payloads and the metal surface. Upon heating by photothermal effect, the energetic barrier for desorption is overcome and the release of active principles takes place. The most used model is the thiol-gold interaction, because of the friendly composition. An example based on small AuNPs (3–5 nm) coated by GSH is reported by Pop and coworkers [106]. GSH plays a critical role in many metabolic pathways, and acts as a protecting agent against free radical oxidation in cells [98]. The controlled release of GSH in specific domains is of interest in metabolic research. In the proposed platform, GSH was desorbed from AuNPs by irradiating with light at the wavelength of the LSPR (in this case around 520 nm) at a power of some watts. Although the reported procedure enables the release in living cells, the use of visible light at relatively high powers is impaired by limited penetration in tissues, severely hampering potential clinical translation. Furthermore, only few drugs provide thiol groups, severely limiting the applicability of direct thiol desorption. It is also important to consider that thiol-coatings are essential for AuNPs stability, and unwanted precipitation or agglomeration of AuNPs can occur upon thiol desorption. On one hand, several approaches that support the release of agents from the coating of the AuNPs rather than directly from the metal surface have been developed, among which are by employing cationic coatings and antisense oligonucleotides.

Delivery of negatively charged residues such as short interfering RNA (siRNA) can be supported by positively charged coatings, such as poly-L-lysine. SiRNA is bound on the surface of coated AuNPs by electrostatic interactions between the negatively charged oligonucleotide and the positively charged polymer coating. Due to the electrostatic interaction, siRNA can be efficiently delivered and released upon irradiation on the LSPR. The temperature increase leads to destabilization of the lysine–oligonucleotide interaction, followed by the localized desorption of the therapeutic payload in the target. The efficiency of the system was tested by silencing the GFP-coding gene by one specific siRNA in living cells, opening exciting perspectives for genetic therapy [107]. Moreover, this approach usually requires low power irradiation, reducing cellular or tissue damages.

In general, the approach of photothermal desorption from AuNPs surface can suffer from some intrinsic limitations, such as the employment of appropriate functional groups on payloads. On the other hand, several drugs or dyes do not possess suitable anchoring groups. Some innovative approaches that bypass this drawback are based on the inclusion of the active agent inside micromaterials containing AuNPs, or inside AuNPs themselves (Figure 3.16). In these synergistic nanomaterials, the release usually results from phase transition of the AuNPs polymeric coating during NIR-irradiation because of the local increase of temperature.

On one hand, an example considers microparticles composed of polymers with a melting point close to 40 °C, the active agent, and AuNPs [108]. The photothermal effect promoted by AuNPs under NIR irradiation triggers the delivery of the drugs because of the melting of the solid matrix. The main disadvantage of this approach is the uncontrolled leakage of the payload from the polymeric matrix. Indeed, defects in the crystallization of the polymeric support can lead to considerable desorption of the drug even without external irradiation. Thermally responsive polymers coupled with porous AuNPs constitute a valid choice for the controlled release of drugs and dyes (Figure 3.16). Indeed, there are some polymer chains that shrink with rising temperature since they undergo a phase transition that increases their hydrophobicity. The shrinkage effect generates pores in the polymeric matrix, allowing molecules to pass through. By exploiting this mechanism, loaded agents can be released by the

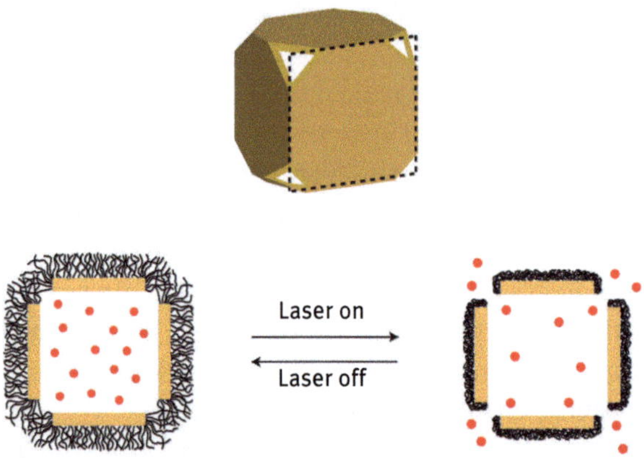

Figure 3.16: Schematic illustration of AuNPs encapsulated with a thermal-responsive polymer. During exposure to NIR laser the light is converted to heat and the polymer collapses, resulting in the escape of the pre-loaded effectors. When the laser is turned off, the polymer chains turn back to the extended conformation and terminate the release. Reproduced with permission from M.S. Yavuz, Y.Y. Cheng, J.Y. Chen, C.M. Cobley, Q. Zhang, M. Rycenga, J.W. Xie, C. Kim, K.H. Song, A.G. Schwartz, L.H.V. Wang and Y.N. Xia, Nature Materials, 2009, 8, 935. ©2009, Nature [78].

hollow or porous nanostructures (Figure 3.16). Among the thermally responsive polymers which have been successfully adopted, it is worth citing poly(*N*-isopropylacrylamide)/polyacrylamide copolymer, which undergoes to phase transition at temperatures tunable in the range 32–50 °C [79]. The peculiar advantage of this approach is the reversible conformational change of the polymer that does not involve the melt of the solid matrix, enabling potential multiple controlled release cycles.

Alternatively, liposomes were also used as thermally triggered carriers for drug delivery. Analogous to copolymers, liposomes can effectively (i) shield the payload and avoid diffusion in the surroundings, while (ii) reversibly shrink by local heating above their phase transition temperature [109].

Unwanted local temperature increases may severely affect the biosafety of all the above nanoplatforms. In this regard, several other approaches for site- and time-specific exogenous-triggered release of active agents have been investigated (Figure 3.17).

The nanoplatforms are usually composed of a caged drug (prodrug) covalently linked to the AuNPs, which may merely act as a carrier. The activity of the drug is suppressed by the covalent linkage, which is usually a photolabile bond that interacts with ultraviolet (UV) light (usually between 250 and 380 nm). For examples, Nakanishi and coworkers (Figure 3.17(a)) proposed a photoresponsive nanocarriers with a primary amine in the structure that photodissociate if irradiated by a pulsed laser at 365 nm with intensity of 100 mW/cm^2 [111]. In addition, Rotello and coworkers developed a gold nanosystem (Figure 3.17(b)) with a photocleavable moiety that dissociates upon UV-light irradiation, resulting in a light-controlled release of the anticancer drug 5-fluorouracil [112]. In their experiments they employed 365 nm

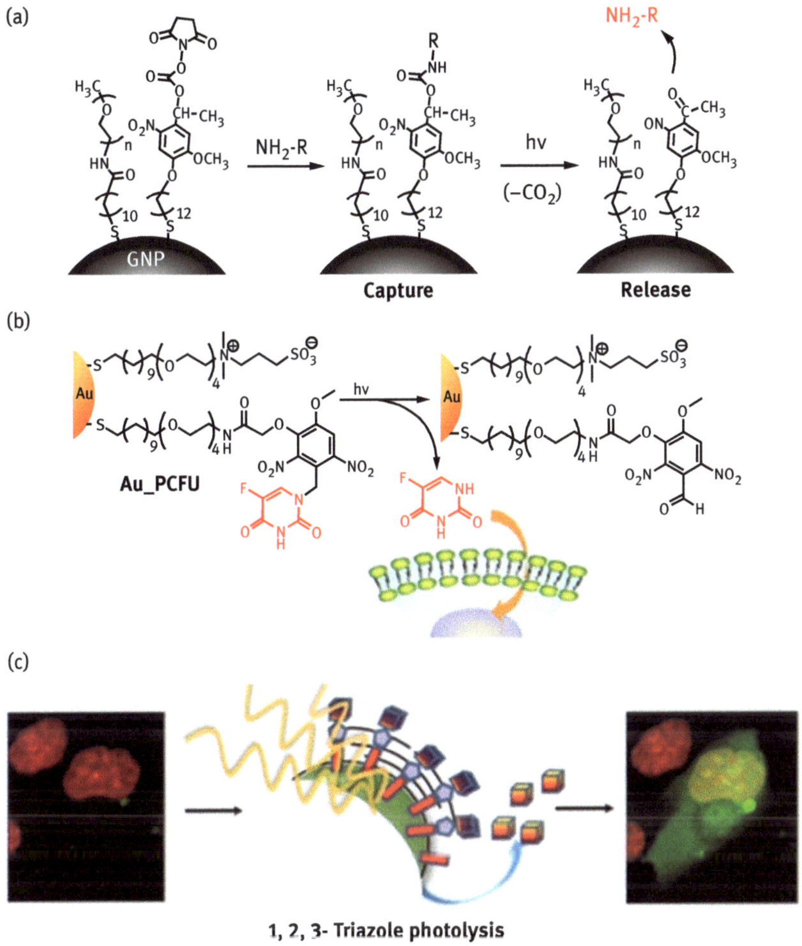

Figure 3.17: Some examples of photoreleasing platforms. (a) Capture and release of amines (NH_2-R) from gold nanoparticles (GNP) having a photocleavable succinimidyl ester. (b) Photochemical reaction at 365 nm of Au_PCFU (gold nanosphere photocleavable 5-fluorouracil) and delivery of the payload to cell. (c) Multiphoton release of fluorescein from gold nanospheres in living cells induced by 561 nm cw laser cleavage of 1,2,3-triazole. Reproduced with permission from V. Voliani, G. Signore, R. Nifosi, F. Ricci, S. Luin, F. Beltram, Recent Patents on Nanomedicine, 2012, 2, 34. ©2012, Bentham Science [110].

laser light for cleaving 5-fluorouracil from AuNPs decorated with *o*-nitrobenzyl groups, and measured the effects of the released drugs by monitoring cell deaths. Despite the good potential of photocleavable nanosystems in medicine, several unsolved problems hinder their translation, including the radiation frequency required for photorelease. It usually belongs to the UV region of the spectrum, resulting toxic and low penetrating in tissues. An elegant approach to shift the irradiation next to the

NIR is by exploiting the optical synergy within metal nanostructures and UV-cleavable molecules [113]. An interesting example is a modular biocompatible nanosystem (Figure 3.17(c)) designed for the photorelease of molecular payloads by irradiation with yellow-green light [99, 114]. These nanoplatforms can effectively release their payload by low-intensity irradiation with cw-laser light at 561 nm, due to the cleavage of UV-absorbing triazole rings [113]. The modular system is composed by four interacting components: (i) 30 nm diameter AuNSs, (ii) peptide coating, (iii) triazolic rings, and (iv) probes to be released [22, 99, 113, 114]. These carriers have been designed to exploit the third harmonic generation of AuNPs [113, 115]. Thanks to this effect, UV-photolabile bonds can be cleaved by irradiating the system at longer wavelengths (visible or near-IR), allowing deeper penetration in tissues. Furthermore, the click-chemistry formation of the 1,2,3-triazole ring linkage between AuNPs and payloads is another advantage of these nanoplatforms, which allows for the friendly composition of the tools. However, further efforts to red-shift the irradiation to the biological-transparency window (800–1,200 nm) are still necessary in order to make these systems more appealing [116].

Overall, the action of nanoplatforms can be triggered by endogenous or exogenous stimuli. The first are usually specific and demonstrate an intrinsic control over the payload delivery, even if they may suffer from unwanted cargo release. The latter are triggered by external stimuli, such as light, which interact with the nanostructures leading to the designed effect. It is interesting to notice that the behaviors of exogenous-triggered nanoplatforms intrinsically reside in the physical and chemical features of AuNPs, as for the light-matter interaction, confirming that AuNPs are pivotal to revolutionize both the diagnosis and treatment of several diseases [8, 113]. Despite several decades of investigations, AuNPs are still not in the market and very few are in clinical trials. In the next section some open challenges for the clinical translation of AuNPs are discussed.

3.3 Challenges and summary

The market of nanomaterials for medicine have increased from 15% in 2014 to 22% in 2019 of the total pharmaceutical market [117]. Nowadays, it relies on organic nanomaterials, and it will increase following the release of noble metal nanoplatforms able to jointly combine synergistic and noninvasive treatments avoiding the issue of metal persistence [2, 6].

Besides metal persistence, AuNPs have to address several other challenges to enhance their possibility of clinical translation, among which are the production protocols that have to be sterile, reasonable, safe, reproducible, and scalable [118, 119]. Moreover, a complete preclinical evaluation of the nanoplatforms, including biosafety and absorption/distribution/metabolism/excretion – ADME package, should be always produced [105, 120]. In this regard, some organizations have been recently emerged to accelerate the transition of nanomaterials to clinical applications. These organizations provide guidelines for the development and characterizations of nanotheranostics supporting researchers with a standardized "Assay Cascade" [6]. Indeed, a peculiar assay cascade is needed for nanomaterials, as their structural complexity makes the regulation guidelines for small molecules somehow inadequate and insufficient. For example, the features of nanomaterials can be easily altered by small modifications in manufacturing processes that can result in limited alterations of the structure but in significant alteration of biological and biodistribution behaviors.

Interestingly, the most underestimated topic is the environmental safety of AuNPs after the medical action [6]. In this regard, the European Commission is focusing on the risks related to the fate of nanomaterials in the air, water, waste, industrial emissions, since 2011, in order to provide further legislative guidance and risk assessment. Indeed, there is still a considerable lack of data and on methods to produce data regarding the exposure of environment and humans to human-made nanomaterials. Consequently, no specific provisions for nanomaterials employment have been yet established by EU legislation (Second Regulatory Review on Nanomaterials COM/2012/0572). The EU Commission is currently working on methods for nanomaterials detection and monitoring and their validation to ensure the proper implementation of the norms.

References

[1] Sun T, Zhang YS, Pang B, Hyun DC, Yang M, Xia Y. Engineered nanoparticles for drug delivery in cancer therapy. Angew Chemie Int Ed 2014;53:12320–12364.

[2] Cassano D, Pocoví-Martínez S, Voliani V. Ultrasmall-in-nano approach: enabling the translation of metal nanomaterials to clinics. Bioconjug Chem 2018;29(1):4–16.

[3] Lim E, Kim T, Paik S, Haam S, Huh Y, Lee K. Nanomaterials for theranostics: recent advances and future challenges. Chem Rev 2015;115(1):327–394.

[4] Cassano D, Voliani V. Behaviors and Persistence of Nanomaterials in Biomedical Applications, Wiley-Scrivener, 2018.

[5] Bobo D, Robinson KJ, Islam J, Thurecht KJ, Corrie SR. Nanoparticle-based medicines: a review of FDA-approved materials and clinical trials to date. Pharm Res 2016;33(10):2373–2387.

[6] Vlamidis Y, Voliani V. Bringing again noble metal nanoparticles to the forefront of cancer therapy. Front Bioeng Biotechnol 2018;6:143.

[7] Cassano D, Mapanao AK, Summa M, et al. Biosafety and biokinetics of noble metals: the impact of their chemical nature. ACS Appl Bio Mater 2019;acsabm.9b00630.

[8] Cassano D, Santi M, D'Autilia F, Mapanao AK, Luin S, Voliani V. Photothermal effect by NIR-responsive excretable ultrasmall-in-nano architectures. Mater Horizons 2019;6(3):531–537.

[9] Chang -C-C, Chen C-P, Lee C-H, Chen C-Y, Lin C-W. Colorimetric detection of human chorionic gonadotropin using catalytic gold nanoparticles and a peptide aptamer. Chem Commun 2014;50(92):14443–14446.

[10] Murphy CJ, Gole AM, Hunyadi SE, et al. Chemical sensing and imaging with metallic nanorods. Chem Commun 2008;(5):544–557.

[11] Chang CC, Chen CP, Wu TH, Yang CH, Lin CW, Chen CY. Gold nanoparticle-based colorimetric strategies for chemical and biological sensing applications. Nanomaterials 2019;9(6):1–24.

[12] Xia F, Zuo X, Yang R, et al. Colorimetric detection of DNA, small molecules, proteins, and ions using unmodified gold nanoparticles and conjugated polyelectrolytes. Proc Natl Acad Sci U S A 2010;107(24):10837–10841.

[13] Zhang L, Salmain M, Liedberg B, Boujday S. Naked eye immunosensing of food biotoxins using gold nanoparticle-antibody bioconjugates. ACS Appl Nano Mater 2019;2(7):4150–4158.

[14] Dykman L, Khlebtsov N. Gold nanoparticles in biomedical applications: recent advances and perspectives. Chem Soc Rev 2012;41(6):2256–2282.

[15] Rosi NL, Giljohann DA, Thaxton CS, Lytton-Jean AKR, Han MS, Mirkin CA. Oligonucleotide-modified gold nanoparticles for intracellular gene regulation. Science (80-) 2006;312(5776):1027–1030.

[16] Medley CD, Smith JE, Tang Z, Wu Y, Bamrungsap S, Tan W. Gold nanoparticle-based colorimetric assay for the direct detection of cancerous cells. Anal Chem 2008;80(4):1067–1072.

[17] Llevot A, Astruc D. Applications of vectorized gold nanoparticles to the diagnosis and therapy of cancer. Chem Soc Rev 2012;41(1):242–257.

[18] Niidome T, Yamagata M, Okamoto Y, et al. PEG-modified gold nanorods with a stealth character for in vivo applications. J Control Release 2006;114(3):343–347.

[19] Kneipp J, Kneipp H, Wittig B, Kneipp K. Novel optical nanosensors for probing and imaging live cells. Nanomed Nanotechnol Biol Med 2010;6(2):214–226.

[20] Guerrini L, Alvarez-Puebla RA. Surface-enhanced raman spectroscopy in cancer diagnosis, prognosis, and monitoring. Cancers (Basel) 2019;11(6):748.

[21] Luin S, Voliani V, Lanza G, et al. Raman study of chromophore states in photochromic fluorescent proteins. J Am Chem Soc 2009;131(1):96–103.

[22] Voliani V, Luin S, Ricci F, Beltram F. Single-step bifunctional coating for selectively conjugable nanoparticles. Nanoscale 2010;2(12):2783–2789.

[23] Cassano D, Rota Martir D, Signore G, Piazza V, Voliani V. Biodegradable hollow silica nanospheres containing gold nanoparticle arrays. Chem Commun 2015;51(49):9939–9941.

[24] Talley CE, Jackson JB, Oubre C, et al. Surface-enhanced Raman scattering from individual au nanoparticles and nanoparticle dimer substrates. Nano Lett 2005;5(8):1569–1574.

[25] Kneipp J, Kneipp H, McLaughlin M, Brown D, Kneipp K. In vivo molecular probing of cellular compartments with gold nanoparticles and nanoaggregates. Nano Lett 2006;6(10):2225–2231.

[26] Alvarez-Puebla R a, Liz-Marzán LM. SERS-based diagnosis and biodetection. Small 2010;6(5):604–610.

[27] Li R, Zhang Y, Tan J, Wan J, Guo J, Wang C. Dual-mode encoded magnetic composite microsphere based on fluorescence reporters and raman probes as covert tag for anticounterfeiting applications. ACS Appl Mater Interfaces 2016;8(14):9384–9394.

[28] Wang Y, Yan B, Chen L. SERS tags: novel optical nanoprobes for bioanalysis. Chem Rev 2013;113(3):1391–1428.

[29] Ruan C, Wang W, Gu B. Single-molecule detection of thionine on aggregated gold nanoparticles by surface enhanced Raman scattering. J Raman Spectrosc 2007;38(5):568–573.

[30] Dreaden EC, Alkilany AM, Huang X, Murphy CJ, El-Sayed MA. The golden age: gold nanoparticles for biomedicine. Chem Soc Rev 2012;41(7):2740–2779.

[31] Sun L, Liu D, Wang Z. Microarray-based kinase inhibition assay by gold nanoparticle probes. Anal Chem 2007;79(2):773–777.

[32] Kim D, Kim K. Controlled synthesis and biomolecular probe application of gold nanoparticles. Micron 2011;42(3):207–227.

[33] Wang X, Qian X, Beitler J, Chen Z. Detection of circulating tumor cells in human peripheral blood using surface-enhanced Raman scattering nanoparticles. Cancer Res 2011;71(5):1526–1532.
[34] Huang X, El-Sayed IH, Qian W, El-Sayed M a. Cancer cells assemble and align gold nanorods conjugated to antibodies to produce highly enhanced, sharp, and polarized surface Raman spectra: a potential cancer diagnostic marker. Nano Lett 2007;7(6):1591–1597.
[35] Qian X, Peng X-H-H, Ansari DO, et al. In vivo tumor targeting and spectroscopic detection with surface-enhanced Raman nanoparticle tags. Nat Biotechnol 2008;26(1):83–90.
[36] Olson AP, Spies KB, Browning AC, Soneral PAG, Lindquist NC. Chemically imaging bacteria with super-resolution SERS on ultra-thin silver substrates. Sci Rep 2017;7(1):9135.
[37] Koh AL, Shachaf CM, Elchuri S, Nolan GP, Sinclair R. Electron microscopy localization and characterization of functionalized composite organic-inorganic SERS nanoparticles on leukemia cells. Ultramicroscopy 2008;109(1):111–121.
[38] Dreaden EC, El-Sayed MA. Detecting and destroying cancer cells in more than one way with noble metals and different confinement properties on the nanoscale. Acc Chem Res 2012;45 (11):1854–1865.
[39] Yang X, Yang M, Pang B, Vara M, Xia Y. Gold nanomaterials at work in biomedicine. Chem Rev 2015;115(19):10410–10488.
[40] Cole LE, Ross RD, Tilley JM, Vargo-Gogola T, Roeder RK. Gold nanoparticles as contrast agents in x-ray imaging and computed tomography. Nanomedicine 2015;10(2):321–341.
[41] Lusic H, Grinstaff MW. X-ray-computed tomography contrast agents. Chem Rev 2013;113 (3):1641–1666.
[42] Chen F, Goel S, Hernandez R, et al. Dynamic positron emission tomography imaging of renal clearable gold nanoparticles. Small 2016;12(20):2775–2782.
[43] Eustis S, el-Sayed MA. Why gold nanoparticles are more precious than pretty gold: noble metal surface plasmon resonance and its enhancement of the radiative and nonradiative properties of nanocrystals of different shapes. Chem Soc Rev 2006;35(3):209–217.
[44] El-Sayed IH, Huang XH, El-Sayed MA. Surface plasmon resonance scattering and absorption of anti-EGFR antibody conjugated gold nanoparticles in cancer diagnostics: applications in oral cancer. Nano Lett 2005;5(5):829–834.
[45] Sonnichsen C, Reinhard BM, Liphardt J, et al. A molecular ruler based on plasmon coupling of single gold and silver nanoparticles. Nat Biotechnol 2005;23(6):741–745.
[46] Kumar S, Harrison N, Richards-Kortum R, Sokolov K. Plasmonic nanosensors for imaging intracellular biomarkers in live cells. Nano Lett 2007;7(5):1338–1343.
[47] Sokolov K, Follen M, Aaron J, Pavlova I. Real-time vital optical imaging of precancer using anti-epidermal growth factor receptor antibodies conjugated to gold nanoparticles. Cancer Res 2003;63(9):1999–2004.
[48] Huang X, Neretina S, El-Sayed MA. Gold nanorods: from synthesis and properties to biological and biomedical applications. Adv Mater 2009;21(48):4880–4910.
[49] Aaron J, Travis K, Harrison N, Sokolov K. Dynamic imaging of molecular assemblies in live cells based on nanoparticle plasmon resonance coupling. Nano Lett 2009;9(10):3612–3618.
[50] Durr NJ, Larson T, Smith DK, Korgel B a, Sokolov K, Ben-Yakar A. Two-photon luminescence imaging of cancer cells using molecularly targeted gold nanorods. Nano Lett 2007;7(4):941–945.
[51] Wu X, Ming T, Wang X, Wang P, Wang J, Chen J. High-photoluminescence-yield gold nanocubes: for cell imaging and photothermal therapy. ACS Nano 2010;4(1):113–120.
[52] Yao Y, Zhang N, Liu X, et al. A novel fast photothermal therapy using hot spots of gold nanorods for malignant melanoma cells. Nanomaterials 2018;8(11):880.

[53] Tong L, Wei Q, Wei A, Cheng J-X. Gold nanorods as contrast agents for biological imaging: optical properties, surface conjugation and photothermal effects. Photochem Photobiol 2009;85(1):21–32.
[54]. Tong L, He W, Zhang Y, Zheng W, Cheng J-X. Visualizing systemic clearance and cellular level biodistribution of gold nanorods by intrinsic two-photon luminescence. Langmuir 2009;25 (21):12454–12459.
[55]. He W, Wang H, Hartmann LC, Cheng J-X, Low PS. In vivo quantitation of rare circulating tumor cells by multiphoton intravital flow cytometry. Proc Natl Acad Sci 2007;104(28):11760–11765.
[56] Steinberg I, Huland DM, Vermesh O, Frostig HE, Tummers WS, Gambhir SS. Photoacoustic clinical imaging. Photoacoustics 2019;14:77–98.
[57] Eichler J, Knof J, Lenz H. Measurements on the depth of penetration of light (0.35-1.0 um) in tissue. Radiat Environ Biophys 1977;14(3):239–242.
[58] Armanetti P, Pocoví-Martínez S, Flori A, et al. Dual photoacoustic/ultrasound multi-parametric imaging from passion fruit-like nano-architectures. Nanomed Nanotechnol Biol Med 2018;14(6):1787–1795.
[59] Avigo C, Cassano D, Kusmic C, Voliani V, Menichetti L. Enhanced photoacoustic signal of passion fruit-like nanoarchitectures in a biological environment. J Phys Chem C 2017;121 (12):6955–6961.
[60] Mallidi S, Larson T, Tam J, et al. Multiwavelength photoacoustic imaging and plasmon resonance coupling of gold nanoparticles for selective detection of cancer. Nano Lett 2009;9 (8):2825–2831.
[61] Cassano D, Summa M, Pocoví-Martínez S, et al. Biodegradable ultrasmall-in-nano gold architectures: mid-period in vivo distribution and excretion assessment. Part Part Syst Charact 2019;36(2):1800464.
[62] Le Guével X, Hötzer B, Jung G, Hollemeyer K, Trouillet V, Schneider M. Formation of fluorescent metal (Au, Ag) nanoclusters capped in bovine serum albumin followed by fluorescence and spectroscopy. J Phys Chem C 2011;115(22):10955–10963.
[63] Le Guével X, Hötzer B, Jung G, Schneider M, Le Guevel X, Hotzer B. NIR-emitting fluorescent gold nanoclusters doped in silica nanoparticles. J Mater Chem 2011;21(9):2974.
[64] Huang -C-C, Chiang C-K, Lin Z-H, Lee K-H, Chang H-T. Bioconjugated gold nanodots and nanoparticles for protein assays based on photoluminescence quenching. Anal Chem 2008;80(5):1497–1504.
[65] Mallory M, Gogineni E, Jones GC, Greer L, Simone CB. Therapeutic hyperthermia: The old, the new, and the upcoming. Crit Rev Oncol Hematol 2016;97:56–64.
[66] Robinson J, Wizenberg M, McCready W. Combined hyperthermia and radiation suggest an alternative to heavy particle therapy for reduced oxygen enhancement ratios. Nature 1974;251:521–522.
[67] Oei AL, Vriend LEM, Crezee J, Franken NAP, Krawczyk PM. Effects of hyperthermia on DNA repair pathways: one treatment to inhibit them all. Radiat Oncol 2015;10(1):165.
[68] Mikheeva S, Barrier M, Little SA, et al. Alterations in gene expression induced in day-9 mouse embryos exposed to hyperthermia (HS) or 4-hydroperoxycyclophosphamide (4CP): Analysis using cDNA microarrays. Toxicol Sci 2004;79(2):345–359.
[69] Furukawa K, Kikusato M, Kamizono T, Yoshida H, Toyomizu M. Possible involvement of mitochondrial reactive oxygen species production in protein degradation induced by heat stress in avian muscle cells. J Poultry Sci 2015.
[70] Welch WI, Suhan JP. Morphological study of the mammalian stress response: Characterization of changes in cytoplasmic organelles, cytoskeleton, and nucleoli, and appearance of intranuclear actin filaments in rat fibroblasts after heat-shock treatment. J Cell Biol 1985;101(4):1198–1211.

[71] Calderwood SK, Theriault JR, Gong J. How is the immune response affected by hyperthermia and heat shock proteins?. Int J Hyperth 2005;21(8):713–716.
[72] Richter K, Haslbeck M, Buchner J. The heat shock response: life on the verge of death. Mol Cell 2010;40(2):253–266.
[73] Liu Y, Zhi X, Yang M, et al. Tumor-triggered drug release from calcium carbonate-encapsulated gold nanostars for near-infrared photodynamic/photothermal combination antitumor therapy. Theranostics 2017;7(6):1650–1662.
[74] Theodorou IG, Jawad ZAR, Jiang Q, et al. Gold nanostar substrates for metal-enhanced fluorescence through the first and second near-infrared windows. Chem Mater 2017;29(16):6916–6926.
[75] Li Volsi A, Scialabba C, Vetri V, Cavallaro G, Licciardi M, Giammona G. Near-infrared light responsive folate targeted gold nanorods for combined photothermal-chemotherapy of osteosarcoma. ACS Appl Mater Interfaces 2017;9(16):14453–14469.
[76] Tsai MF, Chang SHG, Cheng FY, et al. Au nanorod design as light-absorber in the first and second biological near-infrared windows for in vivo photothermal therapy. ACS Nano 2013;7(6):5330–5342.
[77] Liu Y, Bhattarai P, Dai Z, Chen X. Photothermal therapy and photoacoustic imaging via nanotheranostics in fighting cancer. Chem Soc Rev 2019.
[78] Yavuz M, Cheng Y, Chen J, Cobley C. Gold nanocages covered by smart polymers for controlled release with near-infrared light. Nat Mater 2009;8(12):935–939.
[79] Chen J, Yang M, Zhang Q, et al. Gold nanocages: a novel class of multifunctional nanomaterials for theranostic applications. Adv Funct Mater 2010;20(21):3684–3694.
[80] Cobley C, Chen J, Cho E. Gold nanostructures: a class of multifunctional materials for biomedical applications. Chem Soc Rev 2011;40(1):44–56.
[81] Skrabalak SE, Chen J, Au L, Lu X, Li X, Xia Y. Gold nanocages for biomedical applications. Adv Mater 2007;19(20):3177–3184.
[82] Riley RS, Day ES. Gold nanoparticle-mediated photothermal therapy: applications and opportunities for multimodal cancer treatment. Wiley Interdiscip Rev Nanomed Nanobiotechnol 2017;9(4):e1449.
[83] Anselmo AC, Mitragotri S. A review of clinical translation of inorganic nanoparticles. AAPS J 2015;17(5):1041–1054.
[84] Chen F, Cai W. Nanomedicine for targeted photothermal cancer therapy: where are we now?. Nanomedicine 2015;10(1):1–3.
[85] Jiang K, Smith DA, Pinchuk A. Size-dependent photothermal conversion efficiencies of plasmonically heated gold nanoparticles. J Phys Chem C 2013;117(51):27073–27080.
[86] Zhao F, Li X, Li J, et al. Activatable ultrasmall gold nanorods for “off-on” fluorescence imaging-guided photothermal therapy. J Mater Chem B 2017;5(11):2145–2151.
[87] Gordel M, Olesiak-Banska J, Matczyszyn K, Nogues C, Buckle M, Samoc M. Post-synthesis reshaping of gold nanorods using a femtosecond laser. Phys Chem Chem Phys 2014;16(1):71–78.
[88] González-Rubio G, Guerrero-Martínez A, Liz-Marzán LM. Reshaping, fragmentation, and assembly of gold nanoparticles assisted by pulse lasers. Acc Chem Res 2016;49(4):678–686.
[89] Katchinskiy N, Godbout R, Hatef A, Elezzabi AY. Anti-EpCAM gold nanorods and femtosecond laser pulses for targeted lysis of retinoblastoma. Adv Ther 2018;1(1):1800009.
[90] Chen Y-S, Frey W, Kim S, et al. Enhanced thermal stability of silica-coated gold nanorods for photoacoustic imaging and image-guided therapy. Opt Express 2010;18(9):8867.
[91] Cassano D, David J, Luin S, Voliani V. Passion fruit-like nano-architectures: a general synthesis route. Sci Rep 2017;7:43795.

[92] Pocoví-Martínez S, Cassano D, Voliani V. Naked nanoparticles in silica nanocapsules: a versatile family of nanorattle catalysts. ACS Appl Nano Mater 2018;1(4):1836–1840.
[93] d'Amora M, Cassano D, Pocoví-Martínez S, Giordani S, Voliani V. Biodistribution and biocompatibility of passion fruit-like nano-architectures in zebrafish. Nanotoxicology 2018;1–9.
[94] Alphandéry E. A discussion on existing nanomedicine regulation: progress and pitfalls. Appl Mater Today 2019;17:193–205.
[95] Barenholz Y. (Chezy). Doxil® – The first FDA-approved nano-drug: lessons learned. J Control Release 2012;160(2):117–134.
[96] Torchilin VP. Multifunctional nanocarriers. Adv Drug Deliv Rev 2012;64:302–315.
[97] Kim C, Ghosh P, Rotello VM. Multimodal drug delivery using gold nanoparticles. Nanoscale 2009;1(1):61–67.
[98] Cassano D, Santi M, Cappello V, Luin S, Signore G, Voliani V. Biodegradable passion fruit-like nano-architectures as carriers for cisplatin prodrug. Part Part Syst Charact 2016;33 (11):818–824.
[99] Voliani V, Signore G, Vittorio O, et al. Cancer phototherapy in living cells by multiphoton release of doxorubicin from gold nanospheres. J Mater Chem B 2013;1(34):4225.
[100] Ghosh P, Han G, De M, Kim C, Rotello V. Gold nanoparticles in delivery applications. Adv drug Deliv Rev 2008;60(11):1307–1315.
[101] Casciari J. Variations in tumor cell growth rates and metabolism with oxygen concentration, glucose concentration, and extracellular pH. J Cell Physiol 1992;151(2):386–394.
[102] Han G, You -C-C, Kim B-J, et al. Light-regulated release of DNA and its delivery to nuclei by means of photolabile gold nanoparticles. Angew Chem Int Ed Engl 2006;45(19):3165–3169.
[103] Hong R, Han G, Fernández JM, Kim B, Forbes NS, Rotello VM. Glutathione-mediated delivery and release using monolayer protected nanoparticle carriers. J Am Chem Soc 2006;128 (4):1078–1079.
[104] Schneider G, Subr V, Ulbrich K, Decher G. Multifunctional cytotoxic stealth nanoparticles. A model approach with potential for cancer therapy. Nano Lett 2009;9(2):636–642.
[105] Faria M, Björnmalm M, Thurecht KJ, et al. Minimum information reporting in bio–nano experimental literature. Nat Nanotechnol 2018;13(9):777–785.
[106] Pop C. Transport and delivery of glutathione into human cells using gold nanoparticles. US Pat. App 2009:12/616,820.
[107] Huschka R, Barhoumi A, Liu Q, Roth JA, Ji L, Halas NJ. Gene silencing by gold nanoshell-mediated delivery and laser-triggered release of antisense oligonucleotide and siRNA. ACS Nano 2012;6(9):7681–7691.
[108] Moon GD, Choi S-W, Cai X, et al. A new theranostic system based on gold nanocages and phase-change materials with unique features for photoacoustic imaging and controlled release. J Am Chem Soc 2011;133(13):4762–4765.
[109] Paasonen L, Sipilä T, Subrizi A. Gold-embedded photosensitive liposomes for drug delivery: triggering mechanism and intracellular release. J Control Release 2010;147(1):136–143.
[110] Voliani V, Ricci F, Signore G, et al. Smart delivery and controlled drug release with gold nanoparticles. Recent Pat nanomedicine 2012;2(1):34–44.
[111] Nakanishi J, Nakayama H, Shimizu T, et al. Light-regulated activation of cellular signaling by gold nanoparticles that capture and release amines. J Am Chem Soc 2009.
[112] Agasti S, Chompoosor A. Photoregulated release of caged anticancer drugs from gold nanoparticles. J Am Chem Soc 2009;131(16):5728–5729.
[113] Voliani V, Ricci F, Signore G, Nifosì R, Luin S, Beltram F. Multiphoton molecular photorelease in click-chemistry-functionalized gold nanoparticles. Small 2011;7(23):3271–3275.

[114] Voliani V, Ricci F, Luin S, Beltram F. Peptidic coating for gold nanospheres multifunctionalizable with photostable and photolabile moieties. J Mater Chem 2012;22 (29):14487.
[115] Bergamini L, Voliani V, Cappello V, Nifosì R, Corni S. Non-linear optical response by functionalized gold nanospheres: identifying design principles to maximize the molecular photo-release. Nanoscale 2015;7(32):13345–13357.
[116] Hernández-Montoto A, Gorbe M, Llopis-Lorente A, et al. A NIR light-triggered drug delivery system using core–shell gold nanostars–mesoporous silica nanoparticles based on multiphoton absorption photo-dissociation of 2-nitrobenzyl PEG. Chem Commun 2019.
[117] Ragelle H, Danhier F, Préat V, Langer R, Anderson DG. Nanoparticle-based drug delivery systems: a commercial and regulatory outlook as the field matures. Expert Opin Drug Deliv 2017;14(7):851–864.
[118] Zamboni WC, Torchilin V, Patri AK, et al. Best practices in cancer nanotechnology: perspective from NCI nanotechnology alliance. Clin Cancer Res 2012;18(12):3229–3241.
[119] Sainz V, Conniot J, Matos AI, et al. Regulatory aspects on nanomedicines. Biochem Biophys Res Commun 2015;468(3):504–510.
[120] Anchordoquy TJ, Barenholz Y, Boraschi D, et al. Mechanisms and barriers in cancer nanomedicine: addressing challenges, looking for solutions. ACS Nano 2017;11(1):12–18.

Index

https://doi.org/10.1515/9781501511455-004

Bei Fragen zur Produktsicherheit wenden Sie sich bitte an:
If you have any questions regarding product safety,
please contact:

Walter de Gruyter GmbH
Genthiner Straße 13
10785 Berlin
productsafety@degruyterbrill.com